WHAT WE ARE:

The physical basis of consciousness

Izi Stoll

Triskelion Press | Denver Colorado

Published by:
Triskelion Press
Denver Colorado

Copyright © by Elizabeth Stoll. All rights reserved.

No part of this book may be used or reproduced in any manner without written permission from the author, except in the case of brief quotations embodied in critical articles or reviews. For further information or specific usage requests, please send any correspondence to: media@triskelionpress.com

ISBN 978-1-7337931-1-7

Printed in the United States of America.

The Library of Congress has catalogued this print edition as:

Names: Stoll, Elizabeth, 1981- author.
Title: What we are: The physical basis of consciousness
Description: Fourth Edition. | Denver, Colorado: Triskelion Press LLC, 2022. | Includes bibliographic references.
Identifiers: LCCN 2019912188 | ISBN 978-1-7337931-1-7
Subjects: LCSH: Consciousness
LC record available at: https://www.lccn.loc.gov/2019912188

Cover photograph 'Catacombs de Paris' provided by Miguel de los Andes. Used with permission.

Author photograph 'Hudson Hill' provided by Alisha Light. Used with permission.

For Jeri, who taught me that the future
is filled with possibility.

WHAT WE ARE:

The physical basis of consciousness

Izi Stoll

Triskelion Press | Denver Colorado

Table of Contents

Introduction. The mind and the brain..........................1

Section I: Historical Perspectives on Consciousness

Chapter 1. What we know and how we know it...............11

Chapter 2. Dualism...24

Chapter 3. Materialism...30

Chapter 4. Emergence..41

Chapter 5. Physical processes which have been
proposed to underpin consciousness...........................47

Chapter 6. Cognitive processes which have been
proposed to underpin consciousness...........................54

Section II: The Current Impasse

Chapter 7. Evidence for bi-directional interactions
between the body and the mental state........................65

Chapter 8. The possibility of supervenience....................75

Section III: A New Theory

Chapter 9. An intuitive explanation for the emergence of
perceptual awareness and self-awareness......................85

Chapter 10. Information: randomness and meaning..........96

Chapter 11. How neurons encode information..............104

Chapter 12. What neuroscience cannot explain...............115

Chapter 13. Information: probability and uncertainty.......123

Chapter 14. A mechanism for bottom-up processing:
cortical neurons physically create information................136

Chapter 15. A mechanism for top-down processing:
information compression affects neural activity..............142

Chapter 16. The connection between energy expenditure
and information processing..149

Chapter 17. Non-deterministic computation yields
semantical and syntactical statements......................... 156

Chapter 18. Cortical information processing is paired
with representative information content..................... 163

Chapter 19. Cortical information processing is paired
with non-deterministic outcomes............................. 175

Chapter 20. Consciousness means exploring the world
and making use of the knowledge gained..................... 185

Chapter 21. Consciousness as natural computation........ 192

Section IV: Exploring Conifold Theory

Chapter 22. A point-by-point summary of the theory........ 201

Chapter 23. Specific predictions of the theory................ 217

Chapter 24. Compatibility with other theories................228

Section V: Applications and Implications

Chapter 25. The structural and functional requirements
for conscious awareness..239

Chapter 26. The mental reconstruction of reality............ 252

Chapter 27. Knowledge and the concept of the self.........258

Chapter 28. Causation and the arrow of time............... 264

Chapter 29. The origins and benefits of consciousness......272

Chapter 30. How much power do we really have?...........277

Conclusions... 283

References.. 290

Acknowledgements... 306

INTRODUCTION
The mind and the brain

For most of human history, the study of consciousness lay solely in the realm of philosophy. But increasingly, our deepest questions can be addressed directly by science.

In fact, it may be impossible to describe consciousness without describing the physical components from which the phenomenon emerges – the networks of neurons, the neurons themselves, the molecules that make up neurons – as well as the electrochemical and mechanical properties that are characteristic of these systems.

Consider this excerpt from Invisible Cities, by Italo Calvino:

> Marco Polo describes a bridge, stone by stone.
>
> "But which is the stone that supports the bridge?" Kublai Khan asks.
>
> "The bridge is not supported by one stone or another," Marco answers, "but by the line of the arch that they form."

> Kublai Khan remains silent, reflecting. Then he adds: "Why do you speak to me of the stones? It is only the arch that matters to me."
>
> Polo answers: "Without the stones there is no arch."

Consider this passage as a metaphor: The stones are neurons, the individual cells that make up brain tissue. The arch is consciousness, and the pavement which lays across it is our understanding of reality. We can travel across the bridge every day of our lives without ever looking at the stones. But the stones are necessary to form the arch, and the arch is required to hold the pathway. The bridge simply cannot exist without its components.

Let's take a closer look at the bridge and how it is engineered, from its functionality to its structure.

To define 'consciousness', we can say there is a primary (phenomenal) consciousness and a higher-order (access) consciousness [1]. Primary consciousness is being aware, having a rich sensory experience (known as perception) and interacting with the world (known as behavior). All animals have primary consciousness, to varying degrees. Higher-order consciousness is being aware of being aware, having the ability to access perceptual experiences, reflect on them, reason about them, and report them. Both primary and higher-order consciousness have a material basis.

Let's first consider the material basis of primary consciousness: How do we explain perceptual awareness?

Information about the world reaches us through our senses: hearing; seeing; smelling; tasting; feeling pressure, pain, and temperature; balance (knowing where our heads are in relation to the ground) and proprioception (knowing where our limbs are in relation to the rest of the body). The informational content we perceive through the senses is called *qualia*, reflecting the *qualitative* nature of perceptual experience.

This incredible streaming perceptual experience is comprised of information collected from the sensory apparatus. Each of the senses is exquisitely capable of converting a specific category of information from the physical world into a signal our brains can understand. In other words, a conscious awareness of the world is permitted by the structure and function of the nervous system.

Take hearing, for example. The internal components of our ears are wired to transduce physical sound waves into electrical signals. The cochlea, the portion of the inner ear that is shaped like a snail shell, is lined with sensory neurons named *hair cells*, so-called for their tufts of stereocilia which bend in response to sound waves. When the hair on a hair cell is deflected by a sound wave, the movement pulls opens special molecule-sized gates on the hair cell. These gates allow positively-charged ions to flow into the cell. That rush of charged ions is a current flow – an electrical signal. The inside of the cell has a negative voltage, but that voltage increases as positively-charged ions start to flow into the cell. The size of the voltage shift is related to the current flow and the resistance across the cellular membrane. When there is a sufficient flow of positively-charged ions, the voltage of the cell soars and that cell sends a signal to other neurons downstream. Small structures within the neuron, called vesicles, are loaded with neurotransmitters. These vesicles fuse with the membrane and release their contents into the synapse.

This entire process takes only a millisecond, one-thousandth of a second. The neurotransmitters expelled from the pre-synaptic neuron float in the synapse (the space between neurons). One of these molecules may hit a receptor on another cell and stick there. When a neurotransmitter molecule binds to a receptor on a post-synaptic neuron, that receptor opens a gate to allow positively-charged ions inside, and the whole process starts again.

Any neuron receiving impulses from *multiple* upstream sources is prompted to propagate a signal, sending the message onwards by the process described above. That necessary redundancy prevents random error messages from being propagated as true messages. A neuron will take a message seriously only if it has verification from another source. After all, it is highly useful and efficient to be able to ignore false information.

Much information is encoded by sensory neurons. For example, the structure of the inner ear is organized to separate pitch. Due to varying stiffness in the underlying cochlear membrane, each hair cell responds best only to a narrow range of sound waves. Sound waves within a narrow frequency band are encoded by a small set of hair cells, while complex sounds may activate a broad pattern of hair cells. These neurons pass the information onwards, to the auditory cortex and other areas of the brain. The amount of deflection of the hair cell encodes information about volume, by increasing the influx of positively-charged ions. This increases the voltage, which increases the firing rate of the cell. Loudness is therefore encoded in the intensity of the signal. Information on pitch and volume – reflecting the frequency and amplitude of sound waves impinging on the hair cell – is then transmitted to the brain through a combination of neural pathways. If a signal arrives a little later from the right ear than the left ear, the noise is interpreted as coming from the left.

So an awful lot of information – including pitch, volume, and directionality of the noise – is transformed into a useable signal and organized before it even gets to your brain.

Another wonderful example of perceptual awareness is vision. Our eyes are wired to take advantage of different wavelengths of light to gather information about our world through high-acuity color vision. Specialized neurons in the retina only respond to certain wavelengths of light. These neurons can be activated in different combinations, providing vast computational power.

The precise combination of neurons firing electrical signals renders the amazing variety of colors we can see – and limits the variety of colors we can see too. If you have a genetic mutation that renders one of your three photoreceptors ineffective, you will lose the ability to differentiate between some wavelengths of light (a condition called color blindness). Information about the location of specific colors within the visual field is embedded in the electrical signals coming from different parts of the retina, carried by the optic nerve. If your optic nerve is damaged, you may lose sight in the field of vision which corresponds to the electrical signals carried by that part of the optic nerve (a condition called hemianopsia). Light signals received by the eye and carried by the optic nerve are organized by their location of origin within the field of vision, then carried to the visual cortex of the brain. Changes in a signal from a given location over time are interpreted as movement. If you damage some region of the visual cortex in your brain, you may become unable to recognize everyday objects, process motion, or differentiate between faces, depending on the exact area affected.

But these explanations focus on the physical basis of primary consciousness, the neural mechanisms of perceptual awareness. So how does higher-order consciousness arise?

To answer this question, neuroscientists generally reference larger cortical areas present in animals with greater levels of self-awareness, like humans, chimpanzees, dolphins, and elephants. That is, we don't just have connections, we also have connections that connect connections to other connections. That incredible level of complexity seems to be required for higher-order consciousness, although we are still unsure what minimal level of connectivity is sufficient to permit the phenomenon or whether biological material is specifically required to make the connections. But we do know one thing: consciousness is not located in one spot in the brain – it's a global phenomenon that arises from increasingly complex connectivity.

Consciousness is incredibly hard to pin down. But to understand something, scientists have a process.

First, we evaluate *correlation*. For example, areas of the brain light up in functional imaging experiments during tasks that require thought and attention – and critically, different areas are active for different tasks. Mental activity indeed seems to be *correlated* with brain activity.

Next, we evaluate whether a particular component is *required* for the phenomenon to occur. This is a pretty easy one too. Certain areas of the brain, if poked, damaged, or destroyed, cause a person to lose consciousness. Certain drugs which affect the functioning of neurons in the brain can affect the experience of qualia, rendering our senses altered and strange. A great deal of optimization took place over evolutionary time in order for our perception to reflect the world we live in, and a great deal of complex processing in real time is necessary for consciousness to manifest. So we understand a lot about how the brain is *required* for mental activity.

Finally, we must evaluate whether a particular component of the system is *sufficient* for the phenomenon to occur. This is the difficult one. No single part of the brain on its own is sufficient to produce consciousness. It takes many different structures working together to manifest this remarkable phenomenon. The mind essentially emerges from complex, interconnected neural networks – no singular component is *sufficient* to manifest consciousness.

It appears that consciousness is a completely new category of stuff. It is an emergent property of the physical world, but it is not physical matter as we know it. Consciousness is simply a process or phenomenon that arises from global brain activity. An emergent property does not negate materialism – after all, consciousness emerges from material reality and is therefore a property of it.

This perspective asserts that there is no divide between the body and the soul. That is, there never was a fusion of the material and spiritual planes of existence, either at the dawn of humanity or at each individual conception. But critically, the theory of consciousness as an emergent property of the brain also does not take the opposite view, completely negating the existence of consciousness as a real thing by implying that our existence is purely material. This is what separates the theory of 'emergence' from most other contemporary philosophies of the mind.

So how is it possible to reconcile dualism and materialism, these two competing philosophical ideas?

The simple answer is this: A bridge is the arch and the stones that comprise it, but the bridge is also so much more.

Section I
Historical Perspectives on Consciousness

CHAPTER 1
What we know and how we know it

In 1994, a man dressed casually in denim, with his hair tied back in a ponytail, stood up to give a talk at a conference on the philosophy of mind organized at the University of Arizona in Tucson. David Chalmers was about to make a name for himself in the field. With a background in mathematics and cognitive sciences, he had given some thought toward breaking down the problem of consciousness into its component parts. The words he spoke that day, recounted in a paper published the following year, defined the outstanding questions in the field so well, they have become legendary [2]:

"The easy problems of consciousness are those that seem directly susceptible to the standard methods of cognitive science, whereby a phenomenon is explained in terms of computational or neural mechanisms. The hard problems are those that seem to resist those methods.... The really hard problem of consciousness is the problem of *experience*. When we think and perceive, there is a whir of information-processing, but there is also a subjective aspect…. When we see, for example, we *experience* visual sensations: the felt quality of redness, the experience of dark and light, the quality of depth in a visual field. Other experiences go along with perception in different modalities: the sound of a clarinet, the smell of mothballs. Then there are bodily sensations, from pains to orgasms; mental images that are conjured up internally; the felt quality of emotion and the experience of a stream of conscious thought."

In other words, we are beginning to understand the brain quite well – its component parts, from molecules to cells to networks – but we still do not understand the mind, the experience of it all. This mystery is known as 'the Hard Problem'.

To quote the neuroscientist V.S. Ramachandran [3]: "How can a three-pound mass of jelly that you can hold in your palm imagine angels, contemplate the meaning of infinity, and even question its own place in the cosmos?"

We are challenged with a truly difficult problem: we must by necessity study our minds with the only tool available – our own minds. The problem of *what we know* (the field of philosophy called metaphysics) is truly entangled with *how we know it* (the field of philosophy known as epistemology). Untangling the knot between these two puzzles is the first step required to solve either one.

The modern study of consciousness began with John Locke and Rene Descartes, who kicked off the age of reason in the seventeenth century. Locke rejected innate knowledge and character, in favor of developing understanding and judgement via sensory perception and logical reasoning. He wrote [4]:

"Let us then suppose the mind to be, as we say, white paper, void of all characters, without any ideas; how comes it to be furnished? Whence comes it by that vast store, which the busy and boundless fancy of man has painted on it, with an almost endless variety? Whence has it all the materials of reason and knowledge? To this I answer, in one word, from *experience*.... First *our senses*, conversant about particular sensible objects, do *convey into the mind*, several distinct *perceptions* of things.... Secondly, the other fountain, from which experience furnisheth the understanding with ideas, is the *perception of the operations of our own minds* within us."

This insightful theoretical framework of human understanding erased assumptions about the state and contents of the soul and placed responsibility for gaining knowledge solely on human endeavor, via the processes of perception and reflection. Locke's work was foundational to The Enlightenment and the ensuing scientific revolution which continues to this day.

Descartes, for his part, did not wish to attack the problem of consciousness in itself, but felt it necessary to describe the phenomenon as a step toward explicating how we understand the world. This side exercise in metaphysics (the question of what we are) was undertaken to serve his real goal: achieving a coherent theory of epistemology (the question of how we gain knowledge at all).

As they say, one should never put Descartes before the horse. We cannot know what consciousness is until we understand how we gain knowledge about it – but of course, any knowledge about mental processes depend on the operation of the very mental processes under investigation. That creates a sort of chicken-and-egg problem. Our perception is suspect, our logic incomplete. Descartes decided the best way to get around this problem was to avoid metaphysics altogether, at least in the first instance, focusing instead on epistemology. Yet he realized that consciousness played a key role in exploring our world – it makes certain information available to the mind. So he returned to this point in order to explain minimally how mental processes could both limit and facilitate epistemology, making several specific arguments about the nature of the mind [5]:

The first argument centered on the *transparency* of mental processes. Descartes argued that "all of my thoughts are *evident* to me (that is, I am aware of my thoughts) and my thoughts are *incorrigible* (that is, I cannot be mistaken about whether I have a particular thought)." Subsequent thinkers would later call both of these points into question.

The second argument centered on the *reflective nature* of mental processes. Although he wisely separated perception (which focuses on external stimuli) and reflection (which focuses on the internal state), Descartes noticed the inherent subjectivity of all thought processes. There was always an 'I' – with memories, beliefs, biases, intentions, and an existing conceptual framework – doing the thinking. In other words, "Any thought necessarily involves knowledge of myself."

The third argument centered on the *representative nature* of mental processes. Descartes noted that "my thoughts come to me *as if* representing something." Thoughts are not themselves content but instead represent content. They are essentially a model of the world, and indeed the primary method of understanding the world.

Fourthly, Descartes focused on *the separateness of mind and reality*. He doubted that his senses could actually reveal the nature of things in any accurate sense. Since ideas and the things they represent are separate from each other, many times removed through the filter of the senses, Descartes found it impossible to tell what really constituted reality. He wrote:

"Even as I speak, I put the wax by the fire and look: the shape is lost, the size increases; it becomes liquid and hot; you can hardly touch it, and if you strike it, it no longer makes a sound. But does the same wax remain? It must be admitted that it does; no one denies it, no one thinks otherwise. So what was it in the wax that I understood with such distinctness? Evidently none of the features which I arrive at by means of the senses; for whatever came under taste, smell, sight, touch, or hearing has now altered – yet the wax remains... I must therefore conclude that the nature of this piece of wax is in no way revealed by [observation] but is perceived by the mind alone."

This series of logical statements led inevitably to a fifth step: questioning *the very reality of existence*. Given the separateness of the mind and the world, and the potential falseness or incompleteness of sensory information, it became apparent to Descartes that indeed nothing may be real at all. He wrestled with this problem, stating:

"I have convinced myself that there is nothing in the world, no sky, no earth, no minds, no bodies. Does it now follow that I too do not exist? No: if I convinced myself of something, then I certainly existed. But there is a deceiver of supreme power and cunning who is deliberately and constantly deceiving me. In that case I too undoubtedly exist, if he is deceiving me; and let him deceive me as much as he can, he will never bring it about that I am nothing so long as I think that I am something. So after considering everything very thoroughly, I must finally conclude that this proposition, I am, I exist, is necessarily true whenever it is put forward by me or conceived in my mind."

This series of reasoned statements – especially the latter two, where he diverged from Locke – landed Descartes in a huge barrel of pickles. He was trapped by his own logic into believing that nothing really existed at all, except his own thoughts and possibly God as well (by the logic that he would not have come up with the idea of God on his own, so God must exist in reality).

Meanwhile Locke acknowledged that our cognitive processes might be imperfect, but they provided the best tools we have at our disposal to attack the problem. This relatively confident and enormously practical assessment allowed humanity to proceed with investigations into metaphysics, leaving Descartes to wrestle alone in the mental trap he had created for himself.

Ultimately, Descartes' argument against objective reality and for the primacy of the mind was rejected by empirical methods, as we grew a better understanding of how matter (like the wax in the fire) could change its form. The laws of physics, beginning with the formal descriptions of thermodynamics and mechanics by Isaac Newton just a few years later, demonstrated that indeed our senses *could* collect practical information about the world and our reason *could* explicate these observations, with these faculties producing knowledge and building technology that was both reliable and useful in navigating the world around us. Locke was vindicated, and modern science was born.

Philosophers began to drop their epistemological objections and focus on metaphysics, striving to understand ourselves and the world we live in. Science quickly grabbed hold of any testable hypotheses produced by philosophy; results of scientific investigation in turn helped philosophers to hone their theoretical concepts in relation to a constantly-improving knowledge of reality.

In the eighteenth and nineteenth centuries, objections to the existence of absolute reality began to drop away and rigorous scientific methodology became the generally accepted manner of arriving at truth.

Yet some philosophers who accepted this premise nevertheless wanted to retain a belief in God, and therefore objected to science as being able to acquire all types of information. In doing so, they created a problematic ideology which endangers honest investigation into metaphysics to this day. It is worth digressing into this echo chamber of epistemological doubt for a few moments, in order to evaluate the value of absolute truth and objective reality by comparison.

Immanuel Kant, one of the greatest philosophers of the modern era, presented a formidable defense of physics, mathematics, universal reality and absolute truth in his *Critique of Pure Reason*, arguing eloquently for empiricism in the tradition of Locke and against the hopeless, complete skepticism of Descartes [6]. This cleaving to science as a path toward knowledge allowed him to make enormous progress on metaphysics. Yet he hesitated, afraid of where science would lead in regard to such issues as the existence of God, the possibility of immortality, and free will [7]. Kant's antecedent belief system led him to stop himself from exploring questions about the soul with true freedom.

"I have found it necessary to deny knowledge… in order to make room for faith," he wrote. So, while Kant made great strides in developing a metaphysics of space and time, his ability to openly explore the metaphysics of the self were restricted by his faith. To protect it, he limited the exercise of his own brilliance and conformed to the argument that some things were known only through God.

Soren Kierkegaard, a Christian ethicist working some years after Kant, faced similar religious quandaries. While he did not intellectually approach the possibility that science might one day disprove God, immortality, or free will, he did argue vehemently that humanity cannot gain an understanding of *what we need to do* merely by observing *how the world is*. With this logic, Kierkegaard concluded that ethics and the search for meaning in life must remain independent from the scientific search for knowledge [8]. "The crucial thing," he wrote, "Is to find a truth which is truth for me, to find the idea for which I am willing to live and die." This belief found new life in the twentieth century with the advent of post-modernism, which held that anything or nothing could be true. Everything became subjective. Who among us, after all, could say what was right?

This muddled attitude, grounded in a renewal of skepticism, led not only to an inevitable moral relativism but a complete distrust in the possibility of absolute truth or objective reality.

Findings in quantum mechanics and general relativity, which appeared to validate the distrust of universal knowledge and the existence of objective reality, fueled these fires in broader society. The post-modernist movement, which arose from the ashes of classical physics and the existential nihilism resulting from two catastrophic world wars, eventually gained traction on both sides of the political spectrum, culminating in the early twenty-first century trend of celebrating irony and narrative over authenticity and rejecting a shared reality in favor of subjective, individual truths. This intellectual tradition, which inevitably leads to a dizzying anarchy across society, is rooted in the philosophy of existentialism.

John-Paul Sartre was one of the progenitors of this philosophy.

Although he was influenced by Kierkegaard, who sought to find meaning in life, Sartre found the world not very giving in this regard. He wrote in his novel *Nausea* [9]: "Nothing happens while you live. The scenery changes, people come in and go out, that's all. There are no beginnings. Days are tacked on to days without rhyme or reason, an interminable, monotonous addition." With a belief system like this, metaphysics almost did not matter. Why bother understanding the mind or reality or anything at all? Why even bother to continue existing?

Martin Heidegger was another existentialist, but the very opposite of a nihilist, whose ideas on metaphysical philosophy became highly influential in western intellectual discourse. He was also a Nazi. Giving a speech as the Rector of Freiburg University in 1933, Heidegger described the "greatness and glory of this new era" upon the election of Hitler as Chancellor of Germany, and many years later he insisted he had seen "no other alternative" to this murderous political movement [10].

Yet despite his willingness to support the extermination of other people, in his philosophy Heidegger advocated for the human impulse to live a meaningful life, by revealing the self through art and using language to describe the personal act of being. Exercising consciousness, in this view, provided purpose to life. But somehow, in arguing for the human potential to create authentic meaning in life by making the most of the time that we have, Heidegger missed the fact that his political beliefs, put into action, actually prevented other people from doing just that.

By the mid-twentieth century, philosophy itself was in a full existential crisis, despairing that nothing could be known for certain. The human mind was incapable of objective reason or truthful observation; science was a hopeless endeavor; there was no absolute reality; everyone had to identify their own truth. These beliefs filtered into the social sciences and stuck there.

Yet at this critical moment, Maurice Merleau-Ponty brought philosophy back in touch with reality through the timely publication of *Phenomenology of Perception* [11]. In the book, Merleau-Ponty argued for "the primacy of perception" in our efforts to understand the world, placing the physical body at the center of the debate by suggesting it deserves attention as both subject (the one understanding) and object (the one being understood). Usefully, Merleau-Ponty differentiated reality, the bodily perception of reality, and the experience of perception, pointing out that these phenomena were fundamentally independent from each other – and critically, all were real. In a prescient tone, he wrote that there is no such thing as "pure sensation" or, likewise, "an atom of feeling", suggesting there was something irreducible about the conscious experience, which stood apart from the physical processes which allowed the collection of sensory data through the apparatus of the body.

Yet in the decades following the publication of Merleau-Ponty's work, the question of consciousness remained intractable, despite leaps and bounds in the related fields of physics and neuroscience.

As David Chalmers explained fifty years later, the easy problem in neuroscience is understanding how the brain works. The 'Hard Problem' is understanding consciousness – as well as the relationship between the qualitative experience of our minds and the electrical activity pulsing through the soft tissue of the brain. It has become apparent that addressing the Hard Problem will require advances in both metaphysics and epistemology. That stubborn knot still needs to be untied.

To understand our world and ourselves – in fact, to understand how we know things at all – we can employ both philosophical reasoning and scientific experimentation. These twin tools may be imperfect, but they are the best tools we have to discover the truth. And it's worth noting – these tools work well together.

Philosophy can be a singular effort, done in isolation, in which a person states certain axioms and follows the line of reasoning until reaching some conclusion. If the axioms are correct and each logical step is justified, the conclusions must also be correct. Philosophy can also be a two-party effort: logical propositions can be analyzed through argumentation, an approach known as the Socratic Method. This practice, at a heyday when Plato wrote his dialogues between 400 and 350 BC, but still exercised today, is capable of exposing hidden biases, errors in reasoning, or gaps in knowledge. It is a useful device for examining philosophical arguments and flushing out untenable assumptions.

The scientific method is another route toward improving our understanding of the world: another tool in the epistemological toolkit. We define a hypothesis regarding a causal relationship, we specify the variables which could affect the phenomenon, we devise ways to measure the effect, and we run the experiment. The more we learn about the laws of nature, the better we can engineer technology to peer into the workings of ourselves and our world. We can then put this new technology to use in our scientific measurements, improving our precision or even looking at things in a new way. This cyclical process allows us to continually question our assumptions about metaphysics and epistemology – so we can better understand ourselves and our world, and the limitations in perception and processing power which prevent us from fully understanding.

So, we grow to understand our world not only by practicing logical reasoning (first demonstrated in the dialogues of Socrates and his contemporaries) but also by practicing the scientific method (first described by Aristotle and his contemporaries). In fact, both methods are necessary to make progress, by complementing each other. A dialectic or train of thought can drive hypothesis-building, while conversely, hypothesis-testing can either validate or invalidate logical propositions. The knowledge gained by employing the scientific method can then be used to hone logical reasoning, so the process can continue. Both strategies – logic and empiricism – are required to drive progress in understanding our world.

The question of consciousness has now reached a critical juncture. Although many thinkers have striven to describe the nature of perceptual awareness and the feeling of selfhood, no physical basis for these phenomena is known to-date. And while neuroscience has made enormous strides in understanding the neural correlates of mental states and behaviors, an explanatory gap persists. We simply do not know what thought *is*.

The goal of this book is to introduce a new hypothesis describing the nature of consciousness. It is based on existing knowledge about the operation of our brains and the physical laws of the universe, and it builds upon the conceptual framework of previous theories, while adding a lot more mechanistic detail. While I will present evidence that supports this hypothesis, other interpretations cannot be excluded at this time. Yet I hope this exercise in logical reasoning aids the design of experiments which could test the new hypothesis directly. These experiments may either support or disprove my hypothesis as currently stated; but in either case, it is worth evaluating. A new line of thinking may, after all, form the basis of a theoretical framework that advances our understanding of ourselves and our world.

Currently there is no consensus on what the mind is, how it operates, or whether it is capable of interacting with the physical world. Before I propose my own theory, I will back up for a moment and explain existing theories of mind, to evaluate the explanatory power of each conceptual framework.

CHAPTER 2
Dualism

Throughout the history of humanity, many people have believed the mind and the body to be distinct and detachable from each other – intertwined temporarily during the lifetime, separate before and afterwards. And while the body is considered a product of this earth, the breath is considered to be a gift of the divine. This belief is called dualism, because it posits a dual nature of human beings, comprised of both body and soul.

This concept forms the basis of many religious belief systems, with the self considered to be something greater than the body and separable from it. This concept is even considered dogma in many classical religious traditions.

Here is one example from the foundational text of Judaism and Christianity:

"Then the Lord God formed the man of dust from the ground and breathed into his nostrils the breath of life, and the man became a living creature." – Genesis 2:7

This breath, representing the spirit of the living person, is regarded as eternal by those who believe in dualism. It gives life to the body, but can exist separately from the body, and is thought to do so after death, joining its maker in some incorporeal form. Another example, this one from the Quran, the foundational text of Islam, explicitly states there is some non-mortal part of human beings which can survive death:

"He gives life and causes death, and to Him you shall be brought back." – Surah Yunus 10:56

But dualism is not restricted to the Abrahamic religions. Another one of the largest religions in human history, Hinduism, explicitly espouses a belief in the duality of body and spirit:

"Atman (the soul or self) is indeed Brahman (universal truth). It is also identified with the intellect, the mind, and the vital breath, with the eyes and ears, with earth, water, air and sky, with fire and with what is other than fire, with desire and the absence of desire, with anger and the absence of anger, with righteousness and unrighteousness, with everything – it is identified, as is well-known, with what is perceived and what is inferred. As it does and acts, so it becomes: by doing good it becomes good, and by doing evil it becomes evil." – Brihadaranyaka Upanishad 4.4.5

All of these belief systems assert the material basis of the body is but a temporary shelter for the soul – that is, a place where the soul can plug into the world – to feel suffering and joy, to take action which will either prolong suffering or lead to redemption.

In each case, the soul is something eternal or at least something longer-lasting than the body. The foundational texts of these theosophical traditions not only explicitly declare the existence of the soul, but also describe a path for people to follow so their eternal selves can end struggle and reach eternal peace.

Indeed a great deal of religious tradition holds that there is a material/spiritual divide – that our true selves are something superimposed on our bodies, some eternally-surviving presence that is capable of merging with the flesh at birth and separating from it at death. But dualism is not restricted to religious belief. There is a tradition of dualism within secular philosophy as well.

In the late nineteenth century, William James broke ground by establishing psychology as a discipline independent of philosophy. In writing the first textbook on the subject, he articulated the difference between the dual processes of intuition and reason [12]. Intuition, driven by bodily needs, serves us fine in most contexts. Meanwhile, James posited that rational thought is needed for artistic endeavors, design, engineering – and handling any "unprecedented situations".

To elaborate on the contributions of intuition and reason to human action, James articulated five types of decision and the processes by which such decisions were made. The first is reliant upon response to circumstance, "in which the arguments for and against a given course seem gradually and almost insensibly to settle themselves in the mind and to end by leaving a clear balance in favor of one alternative, [which] we then adopt without effort or constraint." The second form involves "letting ourselves drift with a certain indifferent acquiescence in a direction accidentally determined from without, with the conviction that, after all, we might as well stand by this course as by the other." The third form "seems equally accidental, but comes from within, and not from without.... We find ourselves acting, as it were, automatically, and as if by a spontaneous discharge of our nerves.... But so exciting is this sense of motion after our intolerable pent-up state that we eagerly throw ourselves into it." James notes this unpremeditated taking of abrupt action is more frequent in some people than others, although such dramatic turns are surely familiar to us all. The fourth form of decision-making, often motivated by some sobering event, is characterized by a "change of heart" or "awakening of conscience" – leading to "an instant abandonment of the more trivial projects with which we had been dallying, and an instant practical acceptance of the more grim and earnest alternative which till then could not extort our mind's consent." In the fifth type of decision, regardless of whether the facts weigh the result, "we feel, in deciding, as if we ourselves by our own wilful act inclined the beam."

James noted that this final category of decision-making indicated a special case, set apart from the others. Yet, while noting the colossal significance of the possibility of volitional effort, with regard to the question of free will, James intriguingly stated that actual mechanisms of will-power, as distinct from motives, "are not matters that concern us yet."

Indeed, much more thought and effort would be directed to articulating the different types of motivations contained within the human psyche, and how they are encoded within the brain, before the question of will-power could even be formally addressed.

Sigmund Freud built on the ideas of William James and others during the early twentieth century, positing that our selves are ruled by both unconscious and conscious drives [13]. He argued that "in every individual, there is a coherent organization of mental processes" called the *ego*. Meanwhile, "the other part of the mind, into which this entity extends and which behaves as though it were unconscious" is called the *id*.

According to Freud, these two parts of the mind are in constant battle. "The ego represents what we call reason and sanity, in contrast to the id which contains the passions.... The ego has the task of bringing the influence of the external world to bear upon the id and its tendencies, and endeavors to substitute the reality-principle for the pleasure-principle which reigns supreme in the id." Freud's practice of psychoanalysis, also known as talking therapy, aimed to help a person reach into their unconscious and articulate its desires, in order to bring these riling bodily passions under conscious control.

The conceptual framework of Freud, which dominated early psychology, echoed the religious dualists. In each case, there was an angel on one shoulder and a devil on the other – the former patiently encouraging discipline and restraint, while the latter appealed to impulse and desire. The battle between the mind and the body, the ego and the id, was simply a modern way of describing the ancient battle between God and Satan. This sense of wrestling with good and evil, of constantly having positive and negative drives pitted against each other, vying for our attention, has been a pervasive theme in the history of both philosophy and theology. Our very concept of consciousness is entwined with our concept of morality – what we are, what we do, and how we choose our actions seem to be intimately connected with what we feel we are *supposed to do* and what we feel we are *driven to do*.

The dualistic view of metaphysics introduces a problematic

physical and metaphysical separation between two planes of existence. What is the ego that is overpowering the id, bringing it under control? Is it just some other way of talking about an incorporeal spirit? If it is not, then is it instead something merely physical?

The answer to this question has important philosophical implications, as the rejection of dualistic thinking easily leads to denying the actual existence of either matter or thought. Descartes, following his reasoning on the primacy of thought to its logical conclusion, denied matter in favor of stating the only true reality was non-material. Yet this approach has fallen out of fashion. Most modern philosophers interested in the problem of consciousness find it more logical to accept material reality and deny the reality of thought instead.

CHAPTER 3
Materialism

Materialism – which arose in response to a long history of dualism in both eastern and western philosophy – is the complete negation of any spiritual realm. This oppositional view posits that we are physical organizations of matter and nothing more. Best summarized by T.J. Huxley in 1875, these philosophers argue that all animals, including humans, are simply automata. Huxley wrote [14]:

"Volitions do not enter into the chain of causation.... The feeling that we call volition is not the cause of a voluntary act, but the symbol of that state of the brain which is the immediate cause.... Volition is an emotion indicative of physical changes, not a cause of such changes."

Huxley argued, in the aftermath of Charles Darwin's *Origin of Species*, that all of human nature could be extrapolated from understanding the root causes in evolution, development, and circumstance. In the following century, findings from current neuroscience began to abolish the concept of the soul, self, or spirit as something separable from our material body. A belief in the mind as a separable component of ourselves is simply no longer compatible with what we know about the brain. Mental processes are tied to the actions of neurons, and they cease when the actions of neurons cease.

Neurobiology effectively killed off dualism. Reducing the problems of perceptual experience and voluntary behavior to their simplest components and rigorously testing correlation has allowed us to make enormous strides in understanding. However, it has led us to an uncomfortable conclusion: the mind is simply a function of the activity in the brain.

In 1991, the philosopher Daniel Dennett published a book promisingly titled *Consciousness Explained*. In this book, and seventeen other thought-provoking tomes, Dennett argues that consciousness does not exist at all, that our very experience of being is an illusion maintained by an automaton. Our selves, he claimed, are flimsy constructs which last only as long as working memory – we can hold a few salient details about the world, but there is no such thing as qualia or any lasting sense of self. This led some disappointed readers to dub the work *Consciousness Explained Away*. Even Dennett himself argues that only a theory of consciousness which explains the phenomenon in terms of unconscious processes can be valid, stating indeed that "to explain is to explain away" [15]. He champions the idea, now quite popular in neuroscience and philosophy, that thoughts simply do not exist. Our qualitative experience is a complete illusion.

The philosopher John Searle has eloquently argued against this view, pointing out that even having the conversation disproved Dennett's logic [16]. What were people talking about, when they claimed to have personal experiences of a qualitative nature, if these did not exist? How and why would people fake such a thing?

Dennett's response was that there could be competence without comprehension. Our behaviors are simply the results of habits built up over time, combining to form the appearance of a cohesive self. There's no way to tell that experiences reported by people actually occur in the way they describe. It is apparent from this argument that Dennett has never seen a double rainbow or heard the opening notes of Miles Davis' *Kinda Blue* – at the very least, he'll never admit to it.

"I regard his view as self-refuting," contends Searle, "Because it denies the existence of the data which a theory of consciousness is supposed to explain."

But that is the very point. Qualia and subjective experience are not considered data by materialists. They argue that awareness – the feeling of pain, the redness of blood, the sound of a voice, the welling of emotion, the warmth of a blanket – is just a delusion, or some trickery of neural function. The neuroscientist Michael Graziano suggests that consciousness is just a process of directing attention while perceptual awareness is simply a mirage sustained by the computations of the brain [17]. Nick Chater similarly argues the mind is 'flat' – that consciousness is an illusory phenomenon with no basis in reality, equivalent to the process of selective attention at best [18]. This gaslighting approach to cognitive neuroscience does nothing to explain the undeniable phenomenon of perception that we all (presumably) experience. Nevertheless, most neuroscientists today agree with a strictly reductionist view – that perceptual awareness is simply not real. It is instead a mere by-product of neural activity which requires no further explanation.

This view is in dramatic contrast to that of most laypeople, who insist they do experience the wonder of qualitative perception. Most people would say they have thoughts and these thoughts are something categorically distinct from neural activity, even if they accept that thoughts do arise from neural activity. So why have so many neuroscientists come to believe the entire phenomenon of consciousness is an illusion?

The doubt took hold for two reasons. Firstly, the very definition of thought is that it is immaterial. If one believes this world is entirely natural, made of matter and energy, then thought *cannot* exist, being neither of these things. Because the existence of thought is incompatible with the prior belief that our world is comprised of already-defined substances, the very concept of thought must be discarded. The second reason that many neuroscientists have rejected consciousness as a mere illusion is because some experiments actually support this conclusion.

In the 1980s, the neuroscientist Benjamin Libet designed an experiment to test the very existence of free will [19]. The experiment made use of the *Bereitschaftspotential*, or 'readiness potential', a large electrical signal originating from the supplementary motor cortex which appears to precede the conscious decision to perform a spontaneous, volitional act. Libet and his colleagues studied this signal in brain activity, to evaluate the timing and effectiveness of top-down control – that is, the exertion of free will.

Libet and his colleagues instructed the subjects to press a button whenever they felt like doing so. During the experiment, the research team recorded brain activity in the subjects with an electroencephalogram (EEG) and muscle activity in the forearm with an electromyograph (EMG). The EEG measured neural activity, including the distinctive readiness potential, and the EMG measured the initiation of muscle movement. The timings of these two large electrical signals, both related to the action, were compared to the exact time the subject reported feeling the urge to initiate the action, based on the position of a dot moving in a regular motion around a circle like a clock.

This experiment had a surprising result. It demonstrated that reported awareness of initiating a voluntary movement occurs *after* the brain activity guiding movement has been initiated. This result suggests that unconscious neural activity gives rise to the action itself, the awareness of the action occurring, and the feeling of wanting to perform the action.

Given the findings from this experiment, it appears that unconscious processes in the brain are the true initiator of volitional acts, and conscious exertion therefore plays no part in their initiation. If unconscious brain processes instigate some action before consciousness is aware of any desire to perform the action, then there can be no causal role of conscious thought in volition. In the words of Susan Blackmore [20]: "Conscious experience takes some time to build up and is much too slow to be responsible for making things happen."

Therefore, it appears the self cannot be exerting a force of will. Libet suggested that conscious volition may instead be exercised as "the power of veto" or, in the words of V.S. Ramachandran, a sort of "free won't". Perhaps conscious acquiescence plays no role in initiating volitional action, but can actively suppress urges for movement or can permit the unconscious buildup of the readiness potential to be actualized as a movement. Still, this interpretation of the experiment leaves us without free will.

Over the years, however, this intriguing experiment has been shown to have three major problems in its design. Firstly, it relies on the subjective reporting of when a person feels the intention to move. There may be a delay between the impulse to record the clock and the act of recording of it, since this process requires a shift in attention to the clock.

As the neuroscientist Patricia Churchland has pointed out, it may be challenging for the subject to accurately record the moment of decision retrospectively, due to the speed of neurotransmission and the nature of perception [21]. While "antedating" the subjective point of decision-making is required for the experiment, this particular step necessarily frustrates any effort at precision in establishing the relationship between sensation of awareness and the neuronal readiness potential.

Secondly, it is not clear the readiness potential is strictly related to the decision to move. Additional studies have suggested the readiness potential is related more to imagining or expecting a movement, rather than actually initiating the movement [22].

This point was elegantly illustrated in a modified version of the original Libet experiment, in which subjects demonstrated some readiness potential even when they made the decision not to move [23]. This result complicated the key assumptions of the original experiment and its classical interpretation. In another set of experiments, researchers found differing EEG signatures between trials in which subjects *chose* movement versus trials in which they were instructed to move [24]. In yet another version of the experiment, where subjects were asked to press one of two different buttons, the readiness potential manifested seven to ten seconds before the subject reported being aware of the decision to act [25]. These results suggest the readiness potential is not related to the decision to make a specific movement but rather the decision to move in general. The problem in the experimental design, writes the neuroscientist William Klemm [26], is that "the subject knows as soon as one trial is over that another is beginning…. so the pre-movement increased brain activity could actually reflect conscious processing of the 'rules of the game' and the *will* to obey those rules."

Thirdly, Alexander Batthyany has argued persuasively that the original design of the experiment does not in fact test free will [27]. The problem lies in the form of the instruction given by the experimenter: "Let the urge [to move] appear on its own at any time without any pre-planning or concentration on when to act." Batthyany argues the subject has no agency here, since a person cannot passively wait for an urge to occur while at the same time being the one consciously initiating it.

This argument wonderfully echoes William James and his description of the different types of decision, recounted in the previous chapter. Specifically, this experiment simply does not test for the fifth type of decision, the kind that involves the force of will. After all, the original experiment was designed for the subject to hit the button at some point. The choice of action is already decided; it is only the timing that is being tested. In this sense, the experiment is identifying the awareness of a decision of the first or second or third type, where reasoning or external circumstance or a vague welling from within eventually lead the subject to make a movement at some point, without any true force of will.

Perhaps further experiments will be designed to resolve this challenging question – whether we merely become aware of actions, all of which are causally and unconsciously determined, or whether we can, in certain conditions, assert our own will over the physical world. Without clear evidence for the latter, most neuroscientists today assume the former case to be the extent of our capability.

After all, extraordinary claims require extraordinary evidence. If the fifth type of decision described by William James does indeed exist, a mechanism must be provided for the exertion of willpower, in terms of neural operations. No such mechanism has been proposed, so there is no reason to believe we might have such a capability. And so neuroscientists generally believe us to be automatons – simple input/output functions guided by purely deterministic factors.

The relentlessly reductionist approach to identifying the contents of the mind has deep roots in the history of neuroscience, going back nearly 130 years. From the time of Santiago Ramon y Cajal, who established the existence of neurons and worked tirelessly to show how their diverse anatomical features could achieve a uniqueness of function; to David Hubel and Torsten Weisel, who recorded the characteristic electrical signals of individual neurons as they responded to external stimuli in real time; to the researchers currently engaged in the modern science of connectomics, who strive to monitor mental processes through functional imaging of the brain; neuroscience has continually endeavored to explain the very stuff of thought in material terms.

This reductionist approach has led some neuroscientists to posit the seat of consciousness in one specific area of the brain – perhaps one highly connected to other areas, whose function is necessary for sustaining alertness and perception. In particular, the claustrum has been suggested by the eminent scientists

Francis Crick and Christof Koch to be the defined region which organizes the operation of consciousness [28]. This theory is informed by the structure of the nervous system, which suggests the claustrum is well-placed for the job. "It connects to every point of the cortex, bidirectionally," Koch asserts. "In a sense, it acts like the conductor of the cortical symphony." Furthermore, this tiny region of brain tissue seems to physiologically maintain conscious activity. "No abrupt and specific cessation and resumption of consciousness have previously been reported, despite decades of electrically stimulating the forebrain of awake patients in the operating room," Koch points out. "But [this region is] different. Here, consciousness as a whole appeared to be turned off and then on again."

This finding is intriguing. But is identifying a region of the brain that is *required* for consciousness the same as identifying what consciousness *is*? And if this region is indeed responsible for consciousness, what are other regions doing?

Other neuroscientists have theorized the seat of consciousness to be elsewhere in the brain. Particularly, the frontal lobes and the hippocampus have been proposed to direct attention and "stream" conscious episodes across time [29]. In the words of neuroscience researchers James Newman and Anthony Grace: "The hippocampus is the primary recipient of inferotemporal outputs and is known to be the substrate for the consolidation of working memories to long term, episodic memories." Since memories are key to experience and the sense of self, could this region of the brain tie perceptions together and house the very phenomenon of self-awareness?

We do not know the answer yet. But we do know that our mental processes have some physical basis. Whatever thought *is*, neuroscientists have shown overwhelming proof that our lived experiences are encoded in the brain, and that our complex genetic makeup allows that to happen. The soft tissue of the nervous system truly does store our ideas and memories in its physical structure and electrochemical activity. During the first few years of life, enormous modifications take place across the nervous system – many neurons die, while others change their connections. This process of learning and growing continues, albeit at a lower level, during our entire lives. Those physical changes are the mechanism by which we encode our experience, learn about the people and events in our lives, and become entwined with our culture. There are a million immediate (situational), proximate (developmental), and ultimate (evolutionary) influences on who we are and how we act. And the activity in our brains comprises a useful summary of the patterns we have built up.

Yet while neuroscientists have worked to characterize sensation and initiation of action in purely physical terms, an explanatory gap persists in the form of the Hard Problem.

What is this feeling of bound perceptual experience and the unavoidable sense there is an 'I' having the experience? What is the so-called Cartesian theater, the mind's eye, where the perception of reality plays out like a film?

It is possible – the self may be an illusion and we may live in a purely deterministic universe. Maybe the idea of a self as a mere passenger within our bodies is so terrifying that the illusion of control is necessary for sanity, and therefore is under positive pressure for natural selection. Maybe thoughts and feelings are nothing more than brain activity, harmless manifestations of electrochemical turmoil, in no need of explanation.

Alternatively, there may be some yet-undiscovered physical explanation for perceptual experience and some reason such a capacity evolved. There may even be a mechanism for top-down control, whereby our thoughts, arising from neural activity, can subsequently modify that very neural activity to achieve a goal imagined in the conscious realm. The debate continues.

Yet it should be noted: there is room for a sensible compromise here. While the principles of neuroscience will always be dictated by the natural laws of physics, this strict materialism should not preclude consciousness from being considered as an emergent property of a complex system, with corresponding characteristics and rules of operation.

As Benjamin Libet himself once pointed out, "[Consciousness] must simply be regarded as a new fundamental 'given' phenomenon in nature, which is different from other fundamental 'givens', like gravity or electromagnetism." This provocative idea leads us to the theory of consciousness as an emergent property of the brain.

CHAPTER 4
Emergence

Neuroscientists have torn apart the brain, broken it down into its component parts, looked for consciousness everywhere and not discovered the mystery. But is it then logical to conclude that consciousness simply does not exist, or if it does, that it has no causal role in human experience?

"The scientific and philosophical consensus is that there is no nonphysical soul or ego, or at least no evidence for that," the influential philosopher David Chalmers announced at a recent symposium on the subject. The materialistic view is currently very popular indeed. However, this view does not hold great explanatory power. For example, it cannot explain the undeniable phenomenon of streaming perceptual experience.

Chalmers has also carefully pointed out that most non-neuroscientists do accept the existence of streaming perceptual experience, or qualia, and neuroscientists are now "studying the relationship of consciousness to neural and cognitive processes without really trying to reduce it to those processes."

This conceptual framework is called 'emergence' – referring to a process or property arising from a complex system that cannot be predicted or explained by the physical nature of the individual components, or their known interactions with each other. This theory effectively eliminates both classical dualism and pure materialism. It states the mind does arise from the tissue of the brain, yet is not itself material. It is instead a new type of abstraction, not yet fully understood.

The philosopher John Searle has suggested a useful metaphor here, describing consciousness as a function of neural networks, just as digestion is a function of the gut. This constructive terminology should aid scientists in considering the problem as any other in biology.

In fact, this idea is not new. Astoundingly, the concept of the mind as an emergent property of the brain is nearly two thousand years old.

In the second century, the physician-philosopher Galen of Pergamon distinguished between resultant and emergent qualities of wholes – acknowledging that living beings, although built in accordance with the laws of nature, appear to be greater than the sum of their parts [30]. Galen's detailed study of animals, including anatomical dissections and physiological investigations, led to elaborate theories on the localization of function regarding mental and emotional processes – particularly the now-discarded concept of a physical division between rational thinking and impulsive drives, controlled by the brain and the heart, respectively. However, Galen's rigorous scientific approach did lead to heftier ideas that would stand the test of time. In making the argument for a physical basis of psychological processes, Galen posited there was no separation between body and soul – these two were instead intrinsically connected. Yet the mechanism he proposed for this linkage was vague and untenable, with the suggestion that consciousness was birthed through bodily function and swirled through the body to connect the brain and the heart. In his view, the substance of consciousness, called the pneuma, carried the vital and psychic force which animated the body. Although this theory is rather antiquated, it was a critical early step in considering the complexity of the problem of consciousness, and the general relationship between anatomy and physiology.

Yet soon after this work, many of the learned works of antiquity were burned, and it would be nearly 1500 years before the western world re-discovered classical reason and re-established the resources needed to engage in scientific experimentation.

During the Age of Enlightenment, metaphysics and epistemology flourished as a new generation set their minds to the age-old problems of philosophy. After Descartes and Locke, and their contemporaries Leibnitz and Spinoza, attacked the problem of knowledge, other thinkers followed in their footsteps, confidently describing their observations of the world and how human beings used their minds to perceive the world.

One of these next-generation thinkers was Charles Broad. Broad was the first modern thinker to explicitly propose a 'theory of emergence' as the most likely solution to the problem of consciousness [31]. A scientific understanding of the physical properties of the brain was forming, as well as a consensus that mental processes were dependent upon that neural circuitry. However, it appeared to Broad the mind itself amounted to more than the sum of its parts. He wrote: "The characteristic behavior of the whole could not, even in theory, be deduced from the most complete knowledge of the behavior of its components, taken separately or in other combinations." To show the ubiquity of the phenomenon of emergence elsewhere in nature, he presented elegant examples from chemistry: "The characteristic behavior of Common Salt cannot be deduced from the most complete knowledge of the properties of Sodium in isolation; or of Chlorine in isolation; or of other compounds of Sodium, such as Sodium Sulphate, and of other compounds of Chlorine, such as Silver Chloride." The idea of the mind as an emergent property of the brain was born. Yet it was still not clear how the stuff of thought manifested.

Around the same time, in the mid-1920s, the mathematician and philosopher Alfred North Whitehead posed the idea that it is the connection between things that is important, not the things themselves. He wrote in *Science and the Modern World* [32]: "There persists... [a] fixed cosmology which presupposes the ultimate fact of an irreducible brute matter, or material, spread through space in a flux of configurations. In itself such a material is senseless, valueless, purposeless." He argued instead that people and other living things were defined by the process of change. This theory came to be known as process philosophy.

He elaborated on this view in *Process and Reality* [33], arguing there are substances which are actual entities, and there are persistent physical objects, such as human beings, which are abstract entities. The latter, he argued, are best defined as a nexus of temporally overlapping occasions of experience. In this philosophy, the mind is simply a higher level of abstraction from the brain, held together by the continuous existence of function. The primacy of process over substance was a key idea for Whitehead. In accordance with this idea, he wrote that "all things flow" and described a human being as a continuum more than a stable physical form – always changing, with no core identity or essence.

Critically, Whitehead argued that some pairs of processes cannot be connected by simple physical cause-and-effect relations. This idea presents the possibility of an incompletely deterministic universe, leaving space for some exertion of free will. Perhaps somewhere in the zone of thought, unbridled imagination meets the reality-checking influence of sensory perception, and there, decisions are made. Whitehead believed that we are not fully determined by physical causation, as there are some degrees of freedom in our interactions with the world. Contemporary findings in quantum physics heavily influenced his thinking, by suggesting that even electrons had some capacity for creativity – the physical limits of this world still seem to allow some freedom of choice.

Most philosophers and scientists have hesitated to build upon this view, perhaps because we still do not fully understand the physical processes underpinning quantum indeterminacy in nature – much less how these laws might apply to biological systems. Neuroscience as a field retains the assumptions of classical physics, and most neuroscientists consider all neural processes to operate according to deterministic principles.

However, many thinkers who subscribe to the theory of emergence today do implicitly follow in the footsteps of the process philosophers, arguing that consciousness is not located in a single area of the brain, but in the neural activity that links brain regions together. For example, the thalamus and the cortex, which play critical roles in sensory perception and initiation of movement, are connected by extensive bidirectional relays. Interestingly, patterns of activity between these two nodes differ between waking and sleep. The flow is what changes, not the structures themselves. The manner in which these neurons fire and relate to one other in different states – the exact pattern of activation – is key to understanding awareness, the cohesiveness of perceptual experience, and task-specific thinking patterns.

As many researchers around the world study the patterns of electrical activity across the brain under various conditions, other scientists have taken the bold approach of trying to explain consciousness itself in physical terms, specifically hypothesizing how electrical activity within the nervous system might give rise to new emergent properties.

These ideas are highly controversial, but they are worth considering as a step toward building an understanding of the physical basis of consciousness. After all, the goal is to produce a scientific theory of consciousness that accommodates qualia as a valid category of data and explains how free will actually works – ultimately describing the phenomena in question in accordance with the laws of nature, in a rigorous and even quantifiable way. To get started, it is worth evaluating theories that aim to connect mental states with the physical substrate and physical processes of neural tissue.

CHAPTER 5
Physical processes which have been proposed to underpin consciousness

Over recent years, a number of intrepid and interdisciplinary scientists have posited physical mechanisms to explain how consciousness might naturally emerge from neural activity, particularly by applying the principles of quantum mechanics and quantum field theory.

Roger Penrose was one of the first individuals to point out that consciousness and quantum mechanics should be discussed in the same breath, and in relation to each other [34]. While quantum mechanics are generally appreciated to apply only at the atomic scale or smaller, Penrose suggested the phenomena might apply to biological cells as well. His theory relies on the idea that entire atoms or molecules within the human body are subject to these laws, rather than classical mechanics.

The key to understanding quantum processes is to remember that matter comes in discrete packets. A subatomic structure, like an electron, cannot be reduced to smaller components. What's more, such a structure can be fairly considered to exist in an undefined state in the present moment. This guides interactions between photons, electrons, and the like at a quantum level.

A 'coherent state' is achieved when multiple particles have a shared set of possible states, dependent on each other and defined by each other. This 'coherent state', as long as it is maintained, is considered operationally equivalent to quantum entanglement. Any observation, detection, or collision of one particle will also affect the other particles. Such an event will define the characteristics of the particle, including its position and momentum and energy state, and because of the coherence, the other particles will be instantaneously defined as well.

Until they are defined, however, the particles exist in a state of uncertainty, known as a state of quantum superposition. Quantum superposition is when an atom or subatomic particle exists simultaneously in multiple possible states, described by a mathematical wavefunction. Entanglement is a situation when multiple particles exist in superposition, not yet defined – but with the ability to be defined, at least in part, by each other.

Essentially, Penrose proposed the idea that electrons within biological cells are entangled in a quantum superposition state, which then collapses into a single reality. In this theory, the coherent particles are able to resolve into a decoherent state spontaneously – for example, when the energy difference between neighboring matter reaches a critical threshold. This theory relies on physical processes to cause collapse of the coherent state, as opposed to the Copenhagen interpretation of quantum mechanics, which posits that quantum superposition is resolved upon observation or measurement, and the many-worlds interpretation of quantum mechanics, which states that a parallel universe opens to realize each alternative outcome of a quantum superposition. It should be said the occurrence of quantum superposition is not under question, but the applicability of this phenomenon to macro-scale biological cells within the human body is quite controversial.

Stuart Hameroff, a psychologist and anesthesiologist, later added to this hypothesis, focusing on possible quantum-scale changes to the internal structure of the neuron, the molecular scaffolding known as the cytoskeleton. He and Penrose specifically proposed that a momentary electron coherence, guiding a synchronized oscillation of water molecules within the molecular structure of the cytoskeleton, causes an electromagnetic event that allows the quantum tunneling of electrons upon the collapse of the coherent state [35]. Although all cells in the body have a cytoskeleton, this theory proposes that quantum processes manifest thought in the nervous system, yet not in other organs. This hypothesis also directly implies that consciousness arises within neurons, rather than emerging from the connections between neurons, again begging the question of what makes neurons so unique in producing consciousness.

This theory makes several specific predictions, employing the gravitational constant and the Planck constant to predict the rate of electron decoherence. However, independent analysis has suggested that typical timescales for quantum decoherence are orders of magnitude shorter than the timescales of neuronal firing or cytoskeletal dynamics [36]. In other words, the lengths of time and distance across the cytoskeleton suggest it should be considered a classical rather than a quantum system.

Penrose and Hameroff have argued this point in two directions: namely, that quantum states may remain coherent for longer periods of time than previously thought and that neural systems do exhibit a coordinated motion of charged particles at various frequencies.

This theory has raised additional questions from both neuroscientific and mathematical perspectives, which the authors continually strive to address [37]. One major issue is that this theory does not really explain consciousness at all – how do quantum oscillations in the cellular cytoskeleton correspond to a cohesive stream of perceptual experience?

Yet the biggest problem with the theory is simple: there is no reason that consciousness should manifest in neurons, but not other cells of the body – after all, every cell contains a similar cytoskeletal infrastructure, comprised of the same types of molecules obeying the same physical laws. What makes the nervous system special? Or is everyone's liver conscious too?

For these reasons, this theory is not widely accepted. However, the idea behind it is usefully creative and has led to experimental and theoretical work on the possibility of quantum mechanical phenomena occurring in the 'warm, wet, and noisy' environment of the brain.

Although this initial bold hypothesis for a physical basis of consciousness may not be strictly accurate, it will be necessary to describe the phenomenon in terms of physical laws. How is it that neural activity can give rise to a cohesive multisensory perceptual experience, with the self at its center? This is called the binding problem. It is a particularly difficult problem, because there is no one part of the brain where all the sensory information is integrated. Regions of thalamus and cortex that are dedicated to sensory processing are uni-modal – dedicated, for example, to visual or auditory information but not both.

There must be some place that houses the so-called Cartesian theater, where events perceived by the body are played out in the mind in a coordinated, cohesive fashion. Is there one particular location within the brain which seats the Cartesian theater, such as the claustrum, or is the stream of consciousness more difficult to pinpoint than that?

Aiming to address this very question, another hypothesis has been put forth to explain consciousness in physical terms. It is called Conscious Electromagnetic Information (CEMI) theory.

The neurophysiologist Susan Pockett [38], a proponent of CEMI theory, argues that consciousness "is not material in the usually accepted sense, but neither is it some kind of non-physical spook." Rather, she states, it is simply "a local, brain-generated configuration of, or pattern in, the electromagnetic field."

The geneticist Johnjoe McFadden, another proponent of CEMI theory, has specifically proposed that every time a neuron fires an electrical signal, it generates a disturbance in the surrounding electromagnetic field, which creates a representation of the information contained in the neural network [39].

This theory aims to solve the so-called binding problem by stating the entire electromagnetic field generated by the brain unites all of the information encoded in the simultaneous activity of millions of neurons [40]. The theory further explores how extrinsic information entering through the senses must retain some truth about the nature of the represented object or scene, and this 'gestalt' must be extracted from the neural coding.

Criticism of CEMI theory includes the key argument that electromagnetic fields are a common occurrence in nature and have no specific mechanism for manifesting consciousness, and furthermore there is no good way to formally test such a hypothesis. Meanwhile, proponents of the theory argue it is a legitimate line of inquiry, especially as we are only beginning to formulate experimental designs to test it. One key prediction of the theory, that exogenous EM fields should be able to influence neural activity, appears to be supported by a number of studies [41,42]. Moreover, endogenous electrical fields, produced by populations of neurons in a section of neocortex, have been shown to affect the electrochemical activity of neighboring neurons under physiological conditions [43,44]. This theory remains highly controversial, as its proponents explore the scientific and philosophical implications of the theory and consider how it could be more formally tested.

Yet the really intractable problem with this theory is that it posits there is something about EM fields that alone gives rise to consciousness. Again, what makes the nervous system special? Or is an AM/FM radio necessarily conscious too?

Some philosophers and scientists have argued that consciousness may just be a fundamental property of this world, or at least a characteristic of sufficiently complex networks. An extreme version of this view, called pan-psychism, posits that all things have a consciousness and that all things plug into a universal consciousness which experiences the physical sensations, emotions, and thoughts of every living being in the universe. The latter view, in line with many eastern theologies, rests on the assumption that knowledge is a more fundamental substance than matter.

Despite this ongoing philosophical dispute, most people rigorously considering the problem of consciousness today agree the material world is primary and the mind must have some material basis, even if the natural laws governing this phenomenon are not clear yet.

It will inevitably take some work to explain what thought is, and how thought manifests from biological or synthetic networks. The physical explanations which have been posed to-date have been extensively challenged by detractors in the scientific community [45]. Alternative theories of consciousness put forth in recent years have relied instead on describing the phenomenon in terms of cognitive processes.

CHAPTER 6
Cognitive processes which have been proposed to underpin consciousness

The question of consciousness is actually two-fold. The first question is: What is thought? Is it some spiritual substance, as described by the dualists; a non-existent hoax phenomenon, as described by the materialists; some emergent property of interconnected neural networks, as described by the proponents of emergence theory; some coordination or disturbance in the electromagnetic spectrum, as described by the proponents of quantum theories of consciousness? None of these ideas seem fully satisfactory, and none address the second question: What is the self?

Newer theories of emergent consciousness try to explain the phenomenon of thought in terms of the cognitive processes which accompany neural activity. In doing so, they avoid the physical mechanisms underpinning thought completely, but begin to address the question of the self as the central point where thoughts converge.

One could consider the self as the sum of all inputs, from the genes inherited from the parents to the environmental influences occurring in the present and the memory of all remembered events in the past. Gerard Edelman and others have taken this view, suggesting that present neural network function is simply the optimized end state of all these factors converging to most effectively guide advantageous behavior [46].

In contrast to this largely ordered, rational view of neurobiology and behavior is the view of Antonio Damásio, who created the Somatic Marker Hypothesis [47]. This hypothesis suggests instead that emotions and their biological underpinnings – changes to the heart rate or activation of neurons in the amygdala – heavily contribute to decision-making. These bodily cues tend to trigger emotions which influence our behavior, often unconsciously, thereby leading to positive or negative patterns of behavior.

In this view, emotions lay the foundation of consciousness and provide motivation for social behavior. This theory has usefully influenced fields as diverse as neuro-economics, the study of human social interaction, the study of addictive behavior, and moral philosophy.

As in the rationalist view, which posits a cohesive and integrated unit, the thread that connects the hubs of sensory input, genetic tendencies, and stored memory is *the self*. Damásio writes: "The nonconscious neural signaling of an individual organism begets the proto-self, which permits... core consciousness, which allows for an autobiographical self, which permits extended consciousness. At the end of the chain, extended consciousness permits conscience." This concept of a cognitive architecture – grounded in biological function, yet capable of emergent properties, including self-awareness – is a highly-influential descriptor of consciousness today, regardless of whether emotional or rational processes are considered dominant.

These concepts of the self depend upon physical processes. As such, their explanations of the self focus on the physical "I" – the one with *that particular set* of genes and experiences, or the one experiencing *that racing heartbeat*. Yet recently, researchers concerned with the philosophy of mind and the mechanisms underpinning mental states have put less focus on the physical foundations and more emphasis on the cognitive processes that allow perceptual awareness and self-awareness.

Bernard Baars has built a comprehensive conceptual framework called Global Workspace Theory [48]. Building on concepts in cognitive neuroscience, he compares consciousness to a theater – not a Cartesian theater, but an actual theater. There, focal consciousness acts "as a bright spot on the stage, directed there by the spotlight of attention" while "the entire stage of the theater corresponds to working memory". The whole event is produced continuously for the benefit of an audience of one, with unconscious processes working behind the scenes, out of view, off-stage. The holistic experience of the theater provides a singular stream of information, vast yet also intrinsically limited in its representation of reality.

Stanislas Dehaene has extended the Global Workspace Theory with the concept of a "neuronal avalanche" [49]. Information about the world, entering through each sensory apparatus, is integrated by sequential processing steps in thalamic and cortical regions of the brain, then selected by attentional processes. The resulting spotlight of attention, molded by coincident events across sensory modalities, is used to collect additional sensory data which is then broadcast throughout the cortex. In this way, a jumble of inputs across sensory modalities, occurring close together in space and time, can thereby converge onto a single coherent interpretation – for example, a playground filled with screaming children. This singular *interpretation* of objects and events within the global workspace, arising from the functions of perception and memory, thus represents the state of consciousness.

The neuroscientist Michael Gazzaniga has further suggested that conscious events may involve, as a necessary condition, interaction with a "self" system, as a sort of executive interpreter in the brain [50]. This view acknowledges that thoughts arise from neural activity, but argues for the self as a singular, emergent, cohesive – and critically, morally responsible – entity.

Another neuroscientist, Jean-Pierre Changeux, has taken the primacy of the self a step further, arguing that intrinsic mental constructs, not extrinsic reality or sensory input representing extrinsic reality, form the basis of conscious experience [51]. Specifically, he suggests our brains prefer to rehearse existing information sets rather than reacting to externally-derived information, engaging in the latter process only as necessary. Interactions with the environment, rather than being instructive, result merely in the selection of one pre-existing internal representation among many.

Yet regardless of the source of information represented in the brain, it is necessary to articulate how information contained within a given neural network can be integrated to create the bound sensory experience for the 'user'. To this end, the philosopher Giulio Tononi has developed a theoretical unit, called phi, to quantify the level of consciousness in biological and non-biological entities, based upon the complexity of design. This quantity is the amount of integrated information in the whole network and exclusive to the network [52]. Any natural or synthetic entity with computing power can potentially be conceived to have some non-zero level of phi. Both fans of the theory and critics have pointed out that a common household thermostat has some measure of phi, while smartphones and other computers may have a very large measure of phi indeed.

Phi is a useful notion, especially in considering the theoretical and physical requirements for perceptual experience, self-awareness, and other mental functions. This idea is key to the overall conceptual framework of Integrated Information Theory (IIT), as it provides a theoretically measurable quantity of the irreducible information set which is conscious experience [53]. IIT further asserts several axioms to explain consciousness.

IIT posits, firstly, that consciousness is real – that we each have a personal, private experience of thought. Secondly, IIT argues that consciousness is exclusive to the entity experiencing it. Thirdly, IIT posits that consciousness is a conceptual structure, composed of representational objects and events, which form cohesive qualitative experiences – with one experience readily distinguished from another one that contains other objects and other events. Fourthly, IIT posits that momentary consciousness (signified by phi) is irreducible, as individual components of the scene are unified to create a single cohesive experience. And finally, IIT posits the nature of qualitative experience gives rise to the cause-effect structure of our perception of reality. In summary, IIT asserts that consciousness has several key characteristics: it is real, exclusive, distinctive, irreducible, and associated with the perception of cause and effect.

There are three primary criticisms of Integrated Information

Theory. Firstly, the axioms provided are 'self-evident' – not proven from first principles. Secondly, while the principles of existence, exclusivity, irreducibility, the establishment of causal relationships within an information set, the capacity to differentiate between objectively different phenomena, and the capacity to integrate temporally-, spatially-, or categorically-linked subjects are *necessary* conditions for consciousness, these processes are not *sufficient* to produce consciousness. There must be physical parameters that facilitate the manifestation of perceptual experience, but these physical processes or building blocks are not specified in the theory, besides stating they could be either neurons or logic gates.

Thirdly, like other theories describing consciousness in a qualitative manner or as a cognitive process, Integrated Information Theory does not create specific, unique, testable predictions and is therefore unfalsifiable. The few predictions made by the theory do not tell us more about consciousness than we already knew. For example, proponents of the theory cite studies of sleep and anaesthesia which "confirmed the prediction of IIT that the loss and recovery of consciousness is associated with the breakdown and recovery of the capacity for information integration" while other studies confirmed "the differentiation of blood oxygen level-dependent activity patterns decreases when consciousness is lost." [53] While these results are undoubtedly true, they come as no surprise to neuroscientists who already agree that a wakeful brain state is necessary for conscious awareness, and that consciousness is associated with the flow of oxygenated blood to active brain regions. So, while this theory may be useful in articulating the problem of consciousness descriptively, it does not provide any possible mechanism by which consciousness emerges as part of our physical reality, and therefore contributes little to the scientific understanding of the phenomenon. Clearly, a more complete theory is needed.

Indeed, the problem of consciousness was recognized as a grand challenge in the field of psychology at the dawn of the twenty-first century [54]. Addressing the state of the field several years later, researchers had to admit that they still did not know exactly what consciousness was. In considering the question at hand, it had become apparent that "a new openness to interdisciplinary integration of research questions, methods and arguments" would prove "important and fundamental" to addressing this difficult research problem [55]. Indeed, any fully-accountable theory of consciousness will have to be grounded not only in psychological terms, but also in the laws of neuroscience and physics.

While these cognitive theories of mind acknowledge the epiphenomenon of consciousness, none agree on the existence of 'ghostly matter' nor provide mechanistic explanations for thought actually arising from the physical world. Thought is simply, in this view, a cognitive process – generally related to directing attention and constructing a cohesive view of reality. There is no physical mechanism here, only abstract description. And the potential bi-directionality of thought – the possibility for top-down control of neural network activity and the exertion of free will – is not recognized.

So while it is becoming increasingly common for neuroscientists to acknowledge that thought does indeed exist, there is currently no explanation for the phenomenon in terms of physical laws. Furthermore, it is astonishingly rare for researchers and philosophers working in this area to seriously consider thought as participating in the sensory-motor loop. It is of course easier to imagine the brain operating without the additional complexity of non-matter interactions.

Yet while it is possible that consciousness is just some ghostly epiphenomenon along for the ride, it is also possible that thought can actually influence our neural activity and our behavior. This concept is called supervenience.

If the mind can in fact exert causal effects on the brain, participating actively in neural processes – including neurotransmission and initiation of movement – the physical properties and functional operation of mental states might be very different from what we had previously considered. For the purpose of evaluating all possibilities and bringing a fresh perspective to the problem, this idea is worth considering.

After all, we cannot test a hypothesis that we refuse to believe exists.

Section II
The Current Impasse

CHAPTER 7
Evidence for bi-directional interactions between the body and the mental state

Current discoveries in psychiatry and neuroscience suggest the neural tissue of the brain and the ephemera of the mind are connected, and that both are indeed real.

Some contemporary neuroscientists have taken an oppositional view, declaring the mind is simply 'an illusion'. Of course, when it is difficult to understand or explain a phenomenon, sometimes the easiest thing to do is to argue it simply does not exist.

Yet the job of science is not to deny the existence of phenomena, but rather to build theories which explain all the data we have – including the inarguable fact that we do feel conscious thought, emotional states, and subjective perceptual experience – and these are *categorically different* than neural activity.

To understand the natural world and the full range of its wonders, it is necessary to acknowledge the existence of both thought *and* matter. To figure out what they are, without resorting to mysticism or a denial of either phenomenon, is to focus on the physical mechanisms which connect the two.

In the following chapters, I will argue the case for a theory of consciousness that is solidly grounded in neuroscience, but which eliminates both untenable spiritualism and overly-reductive materialism. I will propose that mental processes are something that arises from the physical world – and while consciousness may not be physical matter as we currently understand it, it is a truly physical phenomenon, capable of interacting with matter and effecting causation in the physical world.

It is astonishing that no neuroscientist has seriously explored this idea to date. Indeed, any number of case studies in the annals of neurology demonstrate bidirectional connectivity between the brain and the mind, with each clearly causing changes in the other. In other words, neural network activity in the brain gives rise to conscious experience and that conscious experience appears to affect subsequent neural activity and behavior. Let's consider a few key examples.

Our first case is Phineas Gage, one of the most famous patients in neurology. A railroad foreman, trained to work with explosives, his job was to blast rock to make way for new track. To do this, he used an iron rod to tamp explosive powder into a hole drilled in the rock. One autumn day in 1848, his assistant did not prepare the charge correctly, and as Gage leaned over the hole to plunge the tamping rod, the iron bar projected upwards, passing completely through his head and high into the air [56]. Because the polished conical tip of the tamping rod entered the skull first, it prepared space for the 3 foot long, 1 inch thick, 13 pound cylindrical rod which followed, permitting the brain matter to exit cleanly, without compression of the cerebral tissue in the vicinity of the injury. As a result, Gage had no immediate concussion. He took a carriage to town and sat on the front porch of a hotel until the town physician, Dr John Harlow, arrived. Gage greeted Harlow and proceeded to relate the story of his injury to several curious bystanders as the doctor gazed, astonished, into the man's visibly pulsating brain tissue.

Gage lived for twelve years after the incident. In an age before antibiotics, infection was inevitable, and the days after the accident necessitated close care from the town doctor, who removed stray bone fragments, dressed bandages, and purged the foul-smelling discharge from the wound. Once the infection had passed and the scalp began to heal over, it became apparent that Gage would permanently lose the use of his left eye, but otherwise recover. Harlow noted that Gage remained rational throughout his recovery period. Although moody at times, he had no trouble with sensory perception or making his way about town to do errands; he remembered names and details about his friends; he kept track of the date and the time. These contemporary notes state that he 'does not estimate size or money accurately, though he has memory as perfect as ever."

Psychology was not a science yet, so it must have been a challenge for the town physician to formally articulate certain aspects of mental function. However, the doc did take usefully descriptive notes.

"Is very childish," Harlow wrote about Gage in the weeks following the injury, noting that he had expressed a wish to return to the town of his birth. Later, Harlow would elaborate further [57]: "The equilibrium or balance, so to speak, between his intellectual faculties and animal propensities, seems to have been destroyed. [Gage] is fitful, irreverent, indulging at times in the grossest profanity (which was not previously his custom), manifesting but little deference for his fellows, impatient of restraint or advice when it conflicts with his desires, yet capricious and vacillating, devising many plans of future operations, which are no sooner arranged than they are abandoned in turn for others appearing more feasible."

It is difficult to know the extent or quality of any personality changes, since contemporary mentions of Gage's mental state following his injury often do not corroborate with each other or come from sources familiar with his behavior before the accident [58]. Yet the account of the local physician does agree with a great deal of subsequent neuroscientific findings confirming the role of the frontal lobes in higher cognitive functions, such as reining in impulsive behavior and estimating the size of objects. Before formal neurological studies were conducted to correlate the function of nervous tissue in specific locations of the brain with certain mental states, an accident on a railroad provided a unique and precious insight into the connection between the brain and the personality. The astonishing case history of Phineas Gage tells us a lot about the function and resilience of the human brain.

In the twentieth century, it came to be well-established that neural activity is associated with thought, and generally appreciated that an intact brain is associated with proper mental function. The inverse is true as well: damaging the brain or altering its activity changes mental function.

Indeed, electroshock therapy was once regularly administered to patients with mental illness to stun patterns of brainwaves into submission. This intervention, directed at the ongoing neural activity, was conducted with the express purpose of changing the mental state of the person undergoing the therapy. Patients tended to experience reduction of symptoms after electroshock treatment, along with retrograde amnesia.

In the middle decades of the twentieth century, patients who showed distress or intransigence at being committed to psychiatric care were also sometimes subjected to lobotomy. Clinicians of the time realized that surgical removal of the frontal lobes could take the spark or fight out of a person, and this procedure was performed on patients for this very reason. The problematic thoughts were eradicated by removing the offending brain tissue.

One patient who experienced this medical procedure was Rosemary Kennedy, the sister of President John F Kennedy [59]. Due to a prolonged birth, probably leading to some oxygen deprivation in the brain, Rosemary had intellectual disabilities, reading and writing at only a third- or fourth-grade level. But by all accounts, she was a sweet and vivacious young woman. Her father, however, wished her to be more docile and less moody, so he had Rosemary undergo a frontal lobotomy in 1941, at the age of 23. The procedure left her with the capabilities of a toddler, only able to speak a handful of words and requiring round-the-clock nursing care until her death in 2005, at the age of 86.

For twenty years after the surgery, Rosemary's parents kept her secret. But after the death of their father, the siblings began to visit Rosemary, where she lived under the care of nuns. Eunice Kennedy, in particular, grew close to her elder sister. Their friendship inspired Eunice to campaign on behalf of disabled people across America and worldwide – founding the Special Olympics, advocating for community integration, and working to establish a number of government programs and support services for families with special needs. Her emotional bond with her disabled sister inspired Eunice to dedicate her life to helping others with similar challenges. What happened in Eunice Kennedy's mental sphere translated to the actions she decided to take during her life.

It is apparent from these case studies that neural activity gives rise to thought. The functioning of our brain tissue allows us to walk, talk, think, reason, plan our futures and remember our past. This is not only demonstrated in studies of acute damage caused by traumatic injury or surgical lesion; it is also clear from scientific studies of slower, naturally-occurring processes. Indeed, neurodegenerative diseases – caused by the loss of particular cell types in localized regions of the brain – are associated with characteristic sets of dysfunction in cognitive tasks. For example, Parkinson's Disease is caused by a devastating loss of midbrain neurons which release dopamine into the basal ganglia, and the resulting challenges with movement and motivation have been shown to be reversed by restoring this neurotransmitter into these neural circuits [60,61]. These clinical and laboratory studies prove that damage to the brain can have effects on personality, perception, action, and even the sense of having power over one's movements.

Yet there is a flip-side to this coin. Not only does neural activity affect the content of thought, but it appears that thought in turn can affect neural activity and subsequent behavior. Like Eunice Kennedy, our thoughts can influence our subsequent actions.

The experiences we have and the mental constructs we build – including the personal bonds we forge with other people and the way we imagine the world to be – exist in the realm of thought. But these thoughts can influence the actions we decide to take. Although neuroscientific understanding of such top-down processes is limited, this fact is formally appreciated in medical practice.

One example of top-down control, where thoughts appear to influence neural activity and behavior, involves patients who are paralyzed and thought to be in intractable coma. In recent years, the documentation of some incredible interactions has changed how neurologists treat these patients. These findings suggest that thoughts do exist and top-down control is possible.

Much of this exciting work was conducted by Stephen Laurys and his colleagues at Liege University Hospital in Belgium. This team found that they could coach some patients in a persistent vegetative state to answer yes-no questions when situated in an MRI machine [62]. As a patient willfully generated mental images, blood would flow to those active regions of the brain to provide oxygen and nutrients. This meaningful signal would then be picked up by the researchers. Although the patients, who had often undergone traumatic accidents, showed no outward signs of being conscious, a fraction of these patients demonstrated mental awareness by engaging directly with researchers.

Yet some neuroscientists objected, calling this activity reflexive. The researchers then followed up the initial study using imaging techniques, evaluating whether patients could also manipulate their own brainwaves to communicate with the outside world [63]. They outfitted patients with recording electrodes on the scalp and found that 20% of people previously exhibiting no behavior whatsoever could modify their EEG signal by imagining a specific type of movement, such as curling their toes or swinging a tennis racket, in order to answer yes-no questions about their mental states and preferences. And in yet another study, the researchers aimed to help several completely locked-in patients communicate by changing their pupil size [64]. Completing a mental arithmetic task led to a sympathetic response in the muscles controlling the eye, when the matching answer was heard. This recognition strategy was used to exchange abstract ideas with a locked-in patient without any onerous or invasive equipment. It also proved the paralyzed patients could engage in conceptual thinking and demonstrate behavioral output.

Together, these experiments suggest that mental states can affect bodily function. That is, thought itself appears to have some effect on the subsequent motor output of the individual.

Another excellent example of the power of the mind over the body is the placebo effect. A placebo has no physical effect on the body. It is a pill containing no ingredients, or the context of a surgery without any actual surgical procedure. It is merely an idea planted into the mind, a thought.

The simple idea that one has received a therapeutic intervention – even when none is actually present – causes a significant attenuation of symptoms in a fraction of people [65]. The idea *itself* is sufficient to produce a physiological effect in the body, with patients reporting improvements in symptoms [66] and medical examinations finding changes in objective parameters such as liver enzyme levels [67,68]. The placebo effect is so powerful and so reliable that clinical trials must include placebo groups when evaluating new therapeutic interventions, regardless of the type or route of treatment [69]. The benefit of *merely thinking a medicine is present* is sufficient to produce statistically significant results in a significant number of patients, rendering it necessary for a properly-designed study to take the effect into account.

Placebos can even cause side effects – believing they are taking a medicine, people occasionally derive not only the perceived benefits of the medicine but also the expected side effects [70].

It has been shown that some people are more susceptible to the placebo effect than others; this trait is associated with other personality traits, such as suggestibility [71]. Longer times spent coaching patients on what to expect from the treatment lead to increases in the efficacy of the placebo; furthermore, when the person administering the placebo and the environment of treatment appear more professional, the symptom reduction is noted to be greater [72]. In other words, the more credibly the administration of the placebo is presented as a medicine, the stronger its effect.

Placebos are not anything but a belief. A pill that contains no active chemical formulation cannot have any physical effect on neural pathways – not on temperature sensors or pressure sensors in the skin, not on spinal circuits and cortical regions activated by painful stimuli. Yet empty pills still relieve pain in about a third of patients seeking analgesia. As such, placebos are at the nexus of understanding neuroscience. They are the link between our biological makeup and our psychological experience, and proof that something must link the two. In short, the placebo effect proves that beliefs which exist only in the mental realm can influence bodily function.

In summary, there is abundant evidence for bi-directional interaction between thought and the physical world. The stories of Phineas Gage, Rosemary and Eunice Kennedy, and the millions of people struck with neurological disease every year demonstrate that our brains and our personalities are intrinsically connected. The locked-in patients of Dr Laurys suggest that brain activity and bodily function can be controlled by the imagination. And the remarkable consistency of the placebo effect shows that occurrences in the mental sphere can indeed affect the workings of the body. It is appropriate for neuroscience not to deny these findings but to search for possible underlying mechanisms.

So, let us imagine: not only is the mind real, but it is also a critical component of the sensory-motor loop, exerting causal effects on neural circuitry to affect subsequent behavior.

This argument, arising from and expanding upon emergence theory, allows a complete reassessment of what consciousness is and what it does.

CHAPTER 8
The possibility of supervenience

Friedrich Nietzsche was one the first modern philosophers to approach the possibility that conscious states might be causally efficacious. He argued that consciousness "constitutes a danger to the organism" because it "gives rise to countless mistakes that lead an animal or human being to perish sooner than necessary." This statement implies that consciousness is *actually doing something*. But he did not completely follow this logic that conscious states, when manifested, could influence the actions of the body in the material world – a shame, considering that a human being having such power would abolish nihilism, the other problem that occupied Nietzsche.

While Nietzsche acknowledged that consciousness was not a "necessary" condition for mental processing or goal-directed action, he allowed for the existence of a vast and complex "relation of drives" which affect behavior [73]. The first claim has been borne out by neuroscientific studies of humble organisms like insects: complex and rational behavior can be produced with very simple neuronal circuits which cannot conceivably possess self-awareness. But the latter claim is important too – specifically, the idea that intrinsic drives and emotional states can influence behavior in human beings and other animals. This concept is compatible with the contemporary theories promoted by William James and Sigmund Freud.

Although this view has fallen out of favor in recent years, it is worth considering in detail why that has happened.

The concept that mental processes may have causal efficacy – with thoughts actually influencing events in the physical world – is called supervenience. It supposes that conscious information processing can somehow be transposed to matter and energy. A full theory of supervenience will require the fields of information theory and physics to be combined with cell-level and network-level neuroscience, in order to create a mechanistic explanation for such a phenomenon.

Yet despite major advances in these relevant fields since the nineteenth century, the concept of supervenience has not been greatly explored. Instead, the debate has largely turned toward whether the phenomenon of consciousness exists at all. In the rare occasions that supervenience is broached, it is usually argued against – on the basis that we are essentially zombies whose behavior is completely deterministic.

Someone who has given a great deal of thought to this problem is Jaegwon Kim, a philosopher who works on the theory of mind. He sits firmly in the emergence camp, contending that the qualitative aspect of mental states cannot be reduced to physical matter or processes. "Phenomenal mental properties are not functionally definable and hence functionally irreducible," Kim argues [74]. He suggests instead that consciousness is likely an epiphenomenon, present but incapable of influencing the physical realm.

Kim's argument against supervenience, the causal efficacy of mental states, is based on the principle of causal exclusion, which holds that no event can have more than one sufficient cause. Simply, a behavior cannot have as its cause a physical event and a supervening mental event, without resulting in a case of overdetermination. This would constitute a violation of the principle of causal exclusion. As a result, Kim concludes that physical causes simply exclude mental states from any causal contribution to behavior.

However, this interpretation misses the probabilistic nature of neural function. The world may present some constraints on behavior, but it is not always fully deterministic. Many studies have demonstrated that neural activity at a molecular, cellular, and network level is in fact probabilistic. There remains some degree of freedom here, which allows and may even *require* the intervention of supervening mental states.

Let's consider an example situation.

Imagine that a baseball is flying toward you at 80 miles per hour. The visual information you are currently receiving is useful for coordinating muscle activity in order to take *evasive action*.

Now imagine that you have a mitt on your left hand and memories of playing ball while growing up outside Greenville, South Carolina. You're currently positioned in left field, heart thumping. It's the bottom of the first inning of Game 7 of the World Series. The visual information you are now receiving is useful for coordinating muscle activity to take *effective action*.

Many philosophers, using the principles of reductive materialism, would argue that your prior experience defines your current state. An entity with no qualitative or subjective understanding of the situation, with no supervening power – a perfect zombie – could handle this situation. The task is simple: If you are not trained to catch, not wearing a mitt, not in the context of a baseball game, you duck. If you are a trained athlete, you are wearing a mitt, and you are in the context of a baseball game, you try your damnedest to catch that ball. There is no choice, only a deterministic outcome provided by your neural network training and your contextual cues.

This logic may work for some situations, but not all.

Now imagine that a memory comes to you as you stand in that outfield. Last night, after a heated argument in your hotel room downtown, your teammate Lefty Williams threw $5000 in cash on your nightstand then left the room, assuring you that a number of bookies and their clients knew you had this money and were expecting you to act accordingly. Maybe slow down a bit, throw short to third base, let Cincinnati score a few runs on their home turf and win the series. Keep the money, keep your mouth shut, go back home to Chicago and make love to your wife. All it means is 'forgetting' to be good at baseball for a day or so. This memory sticks in your mind as you stand in the grass at Redland Field in Cincinnati on October 8, 1919 and wonder what to do when that ball comes your way.

Let's call this scenario 'The Eigenstate of Shoeless Joe Jackson'. An eigenstate is a state of uncertainty that remains until it is resolved by some action. A number of behaviors are possible here, and perhaps even equally probable. Shoeless Joe wants to do the right thing, but that extra five grand would double his annual salary. He wants to stay loyal to his team and respect the game, but he also wants to avoid any trouble for himself and his wife, especially with these threats flying around. He has tried to chat with his team owner to find a way to shut down the fix, but Charles Comiskey refused to hold a meeting with him. Shoeless Joe is on his own.

In 1919, the World Series is a nine-game match. After the formidable defense Shoeless Joe and the rest of the White Sox bring to the outfield for Game 7, the series tightens. The match is now 4-3, with Cincinnati only one game ahead. Chicago still has a chance to take the next two games and win the World Series. But the money and the threats of violence are getting persuasive. Meanwhile, there is a great deal of ongoing debate in the ranks – several of the guys are conflicted as to whether they should go through with the fix, some want nothing to do with it, and some are very keen.

Tomorrow is another day and another chance for Shoeless Joe to decide his actions. Tomorrow, the game will be in Chicago. Tonight, he can travel home, discuss everything with his wife, and grow sure in his decision either way.

Ambivalence is uncomfortable, especially when it comes to batting.

Even Shoeless Joe Jackson can go a few games without hitting a home run. He knows no one would find that strange. But then, there's no better feeling in the world than knocking it out of the park. It might even feel better than a few thousand bucks.

The story of the Black Sox Scandal is true, although the inner thoughts of the players are speculation. There's simply no way to know what any of the players were thinking during that time.

However, behavior provides useful information. There's no way a person hits a home run by accident, no matter how strong their training, so Shoeless Joe must have been trying to win Game 8 the following day.

Chicago still lost the game and the series. And as a result of the questionable actions on the field throughout the match, many of the players on the Chicago White Sox that year – including Shoeless Joe Jackson – were implicated in the match-fixing scandal.

However, it's not clear that behaviors were all pre-determined here. While extensive training would have greatly influenced neural activity, it did not require that Shoeless Joe give his full effort at all times. While his early instruction may have inclined him to play a fair game, it may also have inclined him to take action that would keep himself and his wife safe in the face of threats. And while hitting a home run might feel great, getting a pile of money might feel pretty great too. The total amount of dopamine running through the neural pathways of Shoeless Joe Jackson's limbic system – boosted by money and fame and feats of athleticism and winning games, reduced by threats of punishment or threats of harm – could be calculated to measure the net reward of each action. Yet there is no way for that calculation to be made without *information processing*.

I argue that conscious experience and thought are not merely bizarre but useless phenomena – they are instead *mechanisms for information processing*. I argue against dualism, the idea that our bodies and spirits are two entirely separate things – but I also argue against any statement that consciousness is a non-existent or futile epiphenomenon. Both of these theories refuse to acknowledge that current data insists on a more complex reality.

In short, I generally agree with the theory of emergence, but I believe we need to further its explanatory power. To begin, we shall make the following assumptions: that we live in a material reality (e.g. everything in this world has a material origin) and that consciousness is real (e.g. a phenomenon that currently has no mechanistic explanation, yet exists and may even be able to act causally within the neural network that gave rise to it).

Let's take the material world as a given. The complexity of biological organisms evolves, becoming more structurally and physiologically intricate. Eventually, the traits manifested by iteratively connected neural networks add up to something categorically different than what came before. The physical components of this new emergent property cannot be explained with existing terms, and we are forced to build a new language, one that equips us to discuss this new property. These linguistic tools allow us to query the problem further.

Denial of the very existence of qualia – the very fact of conscious perceptual experience – does not help us to understand it. Stating that consciousness is some sort of mysterious mystical stuff does not aid understanding either. Naming the problem does help. Articulating a phenomenon is the critical first step to discovering a mechanistic explanation for it.

I will argue the mind is an emergent feature of neural networks, one that is even capable of causing effects in the physical world through the actions of neurons propagating signals to the body. I contend the mind is a system for information processing – capable of collecting data, storing data, comparing that new data with previously-collected data, and directing action in the world according to the best predictions gleaned from this process, in order to ensure the continued survival of the organism.

Section III
A New Theory

CHAPTER 9
An intuitive explanation for the emergence of perceptual awareness and self-awareness

Although many philosophers have striven to describe the nature of streaming perceptual awareness and the conscious awareness of the self, no mechanism has been proposed that would actually explain either of these phenomena. And while neuroscientists have made enormous strides in understanding the neural correlates of perception, cognition, and decision-making, an explanatory gap persists. We know a lot about the operations of the brain, but we still do not understand what the *mind* is.

I aim to build on these philosophical and scientific foundations by describing a physical basis for consciousness in terms of information processing by a biological substrate. Specifically, I will show how cortical neural networks encode information in a way that is paired with *representational information content*. This theoretical framework builds on two clear premises: firstly, the concept that perceptual experience requires *the integration of sensory information across modalities*, with that information encoded by a neural network; and secondly, the concept that *individual percepts are integrated over time* to build cognitive models. In this view, consciousness allows us to perceive the world, build an understanding of the world, and choose our actions. And it all starts with integrating information!

So, to address what consciousness is, we must first consider *what integration is*. Integration is a mathematical concept which requires summing all things into a single cohesive whole. This fundamental concept was demonstrated by Gottfried Leibnitz and Isaac Newton, working in parallel with each other in the seventeenth century. Yet Greek scholars had, centuries earlier, identified a methodology to approximate the sizes of areas or volumes by breaking the shapes down into components, each described by a finite measurement. For example, the total area under a curved line can be calculated as paper-thin slices of the area under the curve; as the slices get thinner, the accuracy of the measurement improves (Figure 1). Leibnitz and Newton contributed to the field by describing these vanishing increments as infinitely small, thus allowing for a more precise measurement of any continuous mathematical function.

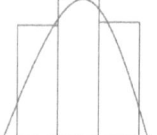 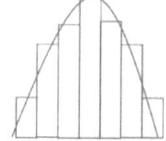  

Figure 1. Approximate integration. Here, the integral of a curved line is measured in an approximate fashion by fitting boxes, each with a defined area, underneath the curve. As the boxes get smaller and smaller, the approximation approaches the true area under the curve. Infinitely vanishing increments, in modern calculus, permit an accurate measurement of this area. Measuring the whole requires *integration* of those component parts.

The calculation of an integral is a fundamental theorem of calculus. The inverse operation is a derivative, which involves calculating a surface description from the whole object. For example, a line can be described as the edge of an area, and a two-dimensional area can be described as the edge, or boundary region, of a three-dimensional volume. With this geometrical approach, it is possible to think of concepts in calculus as concrete concepts in geometry. With this knowledge, we can now explore how dimensions are formed.

We can begin by exploring space in few dimensions (Figure 2). Consider: a point has zero dimensions, mathematically speaking. As it moves through time along a vector, it creates a line. In drawing the line, we have just created a spatial dimension. Now we can imagine this whole line moving through time, and we have created an area. The sum total of space over which the line traveled during a period of time exists in two dimensions – not one, like the line itself. Geometrically speaking, the integration of a line that exists in one dimension (x) is a surface which exists in two dimensions (x, y). Both x and y can change over time to create a two-dimensional geometrical shape – for example, x can change size as the time dimension *becomes* the y dimension, in order to create a shape other than a square. Now this shape moves through time across a third axis. If the shape is and remains a square, its movement creates a cube – the integral of a moving surface that exists in two dimensions (x, y) is a shape which exists in three dimensions (x, y, z).

In other words, time is necessary for creating any spatial dimension and any object changing in time manifests an additional spatial dimension. This exercise is a geometrical form of calculus, and its rule is: integrating an (n)-dimensional shape that is changing over time creates an (n+1)-dimensional shape.

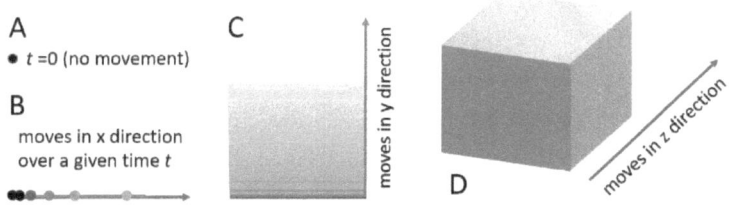

Figure 2. Changes to an (n)-dimensional shape over time create an (n+1)-dimensional shape. (A) A point in zero dimensions is not changing in time. It has only one position on any axis. (B) A point that changes position over time along a single axis creates a line, a one-dimensional shape. (C) A line that changes position over time along a single axis creates an area, a two-dimensional shape. (D) An area that changes position over time along a new axis creates a three-dimensional shape, which takes the form of a cube if the area remains a square during its entire movement. This geometrical law may be extrapolated to higher dimensionality.

Now let's take this logical exercise one step further. The brain is a three-dimensional shape – a complex structure of 86 billion neurons networked together, *which is changing over time*. The integration of neural activity flickering across such a network generates a fourth axis, perpendicular to the others (Figure 3). The resulting shape, described across four axes, is the bound total of all activity occurring in the network.

As the electrical activity across three spatial dimensions (x, y, z) changes, the integration of all this particle movement across the neural network creates a categorically new dimension, which can be called delta (Δ). Delta is the sum total of all electrical activity flickering through this three-dimensional structure over that instant. This integration of neural activity across the entire network *is the integration of all information being encoded*, binding together all information content about the local environment.

The summation of neural activity across the entire brain over a single instant in time yields a cohesive multisensory experience. That neural network state is continually updated from moment to moment, as new sensory input becomes available. Sights and sounds and smells and emotional states come together to provide the impression of a rich perceptual experience.

Imagine the breeze tousling a pink beach umbrella and your position underneath it, looking up, gives a refreshing view of the pink umbrella against the blue sky, combined with a feeling of warmth and the sound of the wind. While the breeze and the umbrella and you yourself exist in the observable universe, your reconstruction of the scene – that is, the perceptual awareness associated with your cortical neural activity – exists in another dimension, defined by the accuracy and completeness of your sensory input.

That rich sensory experience – that flood of information – is encoded by a pattern of neural activity, as the system state changes from one moment to the next. Each instant involves the integration of neural activity across the entire brain. The fourth dimension, delta, is all of the information content encoded by a cortical neural network, which has been physically integrated together. That information content is a multisensory percept.

But now imagine that these individual multisensory percepts are integrated together, creating a fifth axis. The sum total of all those tumbling instants – all that accumulated information – is the self. It is the aggregation of percepts experienced by a single entity over the course of a lifetime *and the one who is experiencing*. This total information set is constantly pummeled with newly-arriving information: snapshots of the neural network state, which can be included or rejected or deemed redundant.

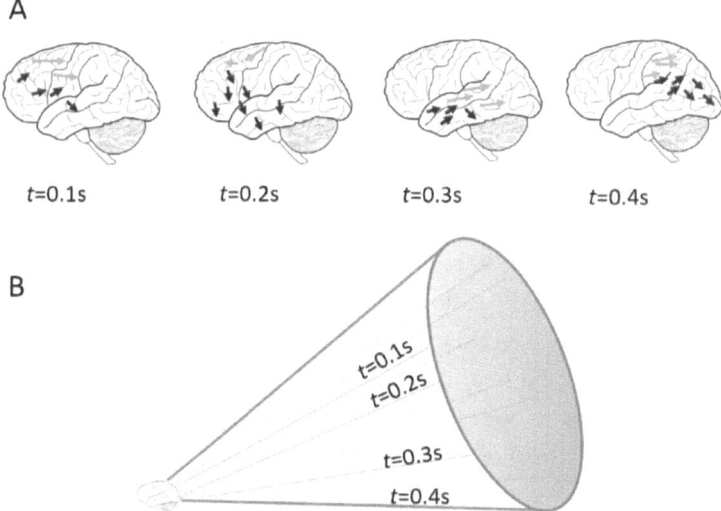

Figure 3. (A) The brain is a three-dimensional structure, and while this structure remains largely intact over small timescales (e.g. seconds to minutes), changes are nevertheless occurring during these timescales (e.g. electrical current flow within the existing structure). Each particle within the structure undergoes some flux; as a result, the three-dimensional structure of the entire brain is *changing* over time. Any instantaneous moment in time can thus be summarized as a vector map for all particles within the brain. (B) The integral of a changing structure necessarily forms a higher-dimensional shape. This schematic depicts a five-dimensional *conifold* rendered in lower-dimensional space. The brain, a three-dimensional structure, is modeled here as a zero-dimensional point. The integrated vector map gives rise to perceptual experience – the information content encoded by the brain in a single moment. This content is contained in a fourth spatial dimension, signified here by one-dimensional lines. The base of the conifold, shown as a two-dimensional surface, is the sum total of all those individual 'snapshots' of neural activity which have coursed through the brain over a lifetime – the integration of all previous neural network states.

If delta is comprised of the flow of information traveling through space, over an instant of time, and is specifically limited to a closed yet interconnected network, the integration of all these deltas is the entirety of information collected by that network. We can call this dimension omega (Ω).

Consciousness, here, is considered a high-dimensional shape. The integration of neural activity across the entire human brain (Δ) is consistent with a cohesive experience of all simultaneous perceptual information that is simply not part of the observable universe. The integration of those percepts – or four-dimensional system states – will generate a five-dimensional shape over time. The integration of those individual system states is the total *information* held by the system. In this view, the total amount of useful (non-rejected) information accumulates over a lifetime. That accumulated information (Ω) is a model of the entire system, the summation of every moment lived. It joins together every immediate (situational), proximate (developmental), and ultimate (evolutionary) influence on the state of being. Omega, here, is the integrated sense of self.

And so, I propose that our bound sensory experience is the sum total of all neural activity across the brain, which mathematically manifests as an additional dimension, describing the system state. The information in this dimension is continuously updated over time, as new information arrives – the denotation Δ signifies change. Every value of delta, every system state, can be summed together, creating an additional dimension that encompasses the entirety of information stored by the brain. The Greek letter Ω is often used to denote the ultimate limit of a set. As such, omega is an appropriate name for that fifth dimension, which sums the entire information content acquired by the system over the course of its existence. Neither dimension is part of the observable universe, but both are part of our physical reality.

The question arises when the additional dimensions may form. Because they are tied to the individual, they cannot exist before the individual exists as an entity, and because the additional dimensions are tied to information acquisition from the world, they cannot exist before the individual begins perception – or more specifically, when the individual begins binding together multiple sensory modalities.

It is therefore a reasonable conjecture that delta begins to flow immediately after birth, when a multisensory experience becomes possible – when sound waves hitting the baby's cochlea are transduced into electrical signals and processed as hearing, photons of light striking the baby's retina are transduced into electrical signals and processed as seeing, being subjected to gravity gives the first sense of balance, being freed of amniotic fluid gives the first sense of proprioception, clearing the nostrils provides the first sense of smell, receiving nourishment through the mouth provides the first sense of taste, and the sudden feeling of air pressure and temperature, which accompany birth, create an individuality of experience and a flow of information. Any organism which integrates multiple sensory modalities could therefore theoretically exist in this additional dimension.

Since multisensory information processing begins at birth, an integrated perceptual experience of these data is therefore also possible at birth. The question then arises when the sense of self forms. This theory asserts that experiences must be integrated over time to build the concept of the self. In agreement with this view, psychological research has shown that young children do not recognize themselves as first-person entities [75]. A cohesive self-perception only appears around age four to five, when the individual begins to express a reliable personal identity and preferences [76], as well as an understanding that others may not hold the same knowledge set as themselves [77,78].

Interestingly, around the same time, all memories from early childhood seem to disappear, with only one or two experiences remaining strongly accessible [79,80]. From a mathematical standpoint, this latter phenomenon is quite interesting – when an integral is integrated, early data is counted many more times than recent data, so entirely removing the early dataset once it has been summarized in a short recent interval allows newer information to weigh more into the dataset. Since the early data is also highly error-prone, its removal in favor of a relatively optimized end state is quite a useful strategy.

The prediction here is that: the information acquired in the years immediately following this data-wipe will be most critical for identity formation, as the data acquired during this period will now be – mathematically speaking – the most represented. If the data-wipe does occur in the manner suggested, the self-concept essentially begins to be constructed from this point in time.

A critical question is whether consciousness can exist separately from the body – that is, whether it is possible for the proposed delta or omega dimensions to persist independently after the encoding structure has ceased to be intact. If consciousness is indeed produced as a function of particle interactions occurring across the dimensions of space and time, then the shape is intrinsically and internally connected. So, it may not be possible for a stream of information to continue to exist without having any continuing input and output. Furthermore, it is not clear that information content would be able to persist without being physically encoded in a lower dimensional structure. There is a neuroscientific argument as well as a mathematical argument: evidence suggests the mind and the brain are not separable or independent. Awareness is dependent on neural activity.

What about the accumulated information in omega? It is unlikely the information itself can exist independently from the neural network that formed it for any meaningful period of time – unless there are physical laws guarding the conservation of information in the universe which we do not yet understand.

What we create during our lifetimes – the words and actions that stick in the minds of other people, affecting their own words and actions – may be the only manner by which we survive after life is over, alongside any genetic contribution we have made to the next generation. Likewise, valuing the things that others have created and the lessons they have learned may be the only way to keep their information in the world.

In summary, biological organisms are temporary arrangements of matter and energy. I will argue that our neural networks are optimized to encode, represent, and process information in a real, physical process of computation – by harnessing the flow of charged particles across the neuronal membrane. This new conceptual framework posits that both perceptual experience (the stream of multisensory awareness) and the concept of the self (the one who is having this experience) are literally the *information content* of the brain, which exists in dimensions only accessible by the system encoding that information content. The delta dimension is awareness itself – the wild flow of sensory perception – bound together to form a cohesive qualitative experience that is not observable to other people. Delta is the integration of all electrical activity across the neural network, and that content is available only to the network that produced it. The omega dimension is formed by the accumulation of these experiences over a lifetime. It is the integrated self, which has certain expectations of the world based on that lived experience.

Simply put, the mind arises from the neural activity of the brain. Streaming perceptual experience is, literally, the integration of information across the neural network over a split second of time. And as the neural network state changes, the mental state changes too. The construct of the self is the sum total of all these moments, integrated together – the sum total of all neural network states and corresponding mental states that we have experienced. Each of us creates a model of the world, based on the information we have collected through our senses, then places the self at the center of this model. The self is that sum total of information, and the one who is collecting further information.

In the following chapters, I will propose a mechanistic process to explain how the brain achieves consciousness. Chapter 10 will introduce some critical concepts in information science. Chapter 11 will cover the principles of classical neuroscience. Chapter 12 will detail where modern neuroscience currently falls short. Chapter 13 will explain how physical systems create information. Chapter 14 will propose a mechanism by which cortical neurons harness probabilistic events to create integrated information, which effectively represents reality. Chapter 15 will explain how information is compressed to extract predictions about reality. Chapters 16-17 will clarify how energy efficiency is related to non-deterministic computation. And finally, Chapters 18-21 will explain how rich perceptual experience and behavioral choice are made possible by biological information processing.

CHAPTER 10
Information: randomness and meaning

We use the term 'information' a lot in everyday life. We talk about 'sensitive information', 'inside information', 'too much information', 'information overload', and 'the information super-highway'. Information generally refers to the transfer or storage of bits of data. Information is considered to be valuable if it carries meaning and considered to be rubbish if it doesn't.

Information is a very real, measurable quantity. For this reason, it requires a certain amount of *bandwidth* to effectively transmit bits of information without losing the integrity of that information. This concept of quantifying information is worth exploring in some detail, so that we can start thinking about information as a quantifiable part of our physical world.

The mathematical definition of information is rather simple. It is the sum of all possible states for a system. For example, a small computer chip may be comprised of two transistors. Each transistor can be in an 'on-state', signified by 1, or an 'off-state', signified by 0. There are four possible states here: 00, 01, 10, or

11. If the system consists of more transistor components, they can be grouped into larger strings of 0s and 1s to represent a large set of symbols, such as letters and numbers. With this binary code, we can use large arrays of transistors – laid out on computer chips – to store, represent, and transfer data.

If there any repeated or predictable patterns in the system, for example if every transistor flips to an on-state if its left neighbor is in an off-state, and flips to an off-state if its left neighbor is in an on-state, then there is little disorder in the system. It is instead a very *ordered* system. By contrast, if the state of each transistor is random, the system is very *disordered*.

So, the fewer the patterns in the dataset, the greater the disorder, and the harder it is to predict other values! This is the case with randomness, where there are many possible combinations of transistor states and therefore a *large quantity of information* from a mathematical standpoint. So, a system with *fewer patterns* or *less predictability* contains *greater amounts of information*.

Yet the everyday definition of information is the very opposite of the mathematical definition. The everyday definition refers implicitly to the amount of *meaning* we derive from a dataset. The more patterns in the system, the more we understand the system, and the better we can predict its true state – in this view, 'information' is equivalent to recognizing patterns in the dataset. With this colloquial definition, a system with *more patterns* or *more predictability* contains *greater amounts of information*.

The problem here is that we are using the same word to describe two different concepts. 'Information' has a strict mathematical definition: the sum of all possible states in a physical system or dataset. 'Information' also has a colloquial definition: the useful patterns in a dataset, or the *predictability* of that dataset. With the mathematical definition, the quantity of information *increases* with meaninglessness, but with the colloquial definition, the quantity of information *decreases* with meaninglessness. As occasionally happens in the English language, one word has developed two opposite meanings. It is not helpful to deny either the mathematical definition or the colloquial one. Both are useful concepts. But before we can think about how we extract *meaning* or *predictability* from a dataset, we have to understand how we generate disorder in the first place.

Appreciating how a system might generate disorder will help us to consider information as a physical quantity. To get there, we must study the history of scientific discovery in this field.

During the nineteenth century, the physicist Ludwig Boltzmann was studying how atoms acted in a gas, in an effort to describe the trajectory of each object across space and time. Boltzmann's research on this problem created the foundation for statistical mechanics – a branch of physics that focuses on randomness.

In the physical world, atoms are always moving around. And the present state of any system comprised of atoms is *the set of possible arrangements of every component particle in that system.*

Boltzmann was considering a gas, trapped in a container with a particular volume and pressure. This physical system, which contains many atoms, has a set of possible *macrostates*, which are defined by the distribution of *microstates*, or the possible states of every component atom. As in our modern transistor example, the state of the entire system is the sum total of possible states for each component part. Boltzmann figured out the key factor: using the natural logarithm of the probability distribution, which accounts for *the time taken* to reach that set of possible macrostates.

And so, Boltzmann created a simple equation, which provides the number of ways that atoms in a gas can be arranged within a thermodynamic system – the number of possible *macrostates*, as a sum of possible *microstates*. He called this quantity *entropy*. The equation was engraved on Boltzmann's tombstone, as a testament to his life's work.

Other researchers built on Boltzmann's efforts. Since microstates within a thermodynamic system are not all equally probable, this complexity must be accounted for. The modified formulation of the law is called Gibb's entropy equation, after Josiah Willard Gibbs [81]. During a lifetime of New England solitude, Gibbs worked to describe the entropy of a thermodynamic system as a discrete set of microstates, each with some probability of occurring. His improved method for calculating entropy not only accounts for the probable state of each component atom in the system and the amount of time that has passed to reach that set of possible states, but also the temperature of the system.

What determines the probability of each microstate? In a gas or a solid structure, the other atoms may provide constraints, simply by occupying a location in space so that it cannot be occupied by another atom. The density of atoms makes a difference too – if atoms are free to move around, but they bump into each other a lot, they will create pressure and heat from these interactions. In these high-pressure, high-temperature systems, the number of possible system states increases dramatically.

The probability of an event is related to how unpredictable or uncertain the state is. If the probability of a certain microstate is high, the overall system will be more ordered and predictable. For example, a perfect crystal structure is very ordered and is therefore simple to describe. To describe the arrangement of the crystal, we first recognize a pattern: the crystal is a stable structure comprised of repeating elements. The probability is high that each atom within a crystal structure has a particular position – its location is likely to be spaced in relation to other atoms and the atom itself is not likely to be moving much. The entire set of likely states for the atom to occupy within a crystal structure is quite small, in comparison with an atom that is situated in a gas, free to move. And so, to describe the crystal, we do not have to write down all the possible states of each atom, or imagine all the possible positions each atom might take within the structure. We can instead guess it will follow a regular pattern, and we write down that chemical structure instead. Because the probabilities are so overwhelmingly in favor of one arrangement, the crystal creates very little entropy – very few possible system states. It does not tend to change over time, and it does not create heat.

A crystal – unchanging across millennia, virtually incapable of any interactions which could give rise to new atomic arrangements – has negligible amounts of entropy. Meanwhile a gas – always changing, with an astounding range of possible microstates – has large amounts of entropy. In other words, a crystal is *more ordered* than a gas. Therefore, it takes fewer *bits of information* to describe the arrangement of atoms in a crystal than in a gas.

Describing the number of possible macrostates for a system – its entropy – is about describing the overall uncertainty of every component microstate. The person who discovered this general rule was a telecoms expert named Claude Shannon.

Shannon lived a generation after Boltzmann and Gibbs. In 1937, he completed his master's degree in digital circuit design theory by showing how electrical circuits encode information. He then worked to establish secure telecommunications during World

War II, participating in codebreaking efforts for the Allies. His work in cryptography – figuring out how to securely transmit information in discrete quantities proportional to the available bandwidth – led to the breakthrough that he is remembered for today: articulating the mathematical definition of information.

Shannon realized the total quantity of information in a system is the sum of all unlikeliness and uncertainty [82]. The more *possible* macrostates a system was capable of having, the more *unlikely* any one macrostate would be. The total information held by the system is the sum of probabilities for all component microstates, multiplied by the natural logarithm of each probability.

Shannon was stunned to recognize the similarity of his own equation to the entropy equations of Boltzmann and Gibbs. In fact, this newly-derived equation describing *information* was identical to equations of *entropy* describing physical particles – but without Boltzmann's gas constant for physical systems.

Both information and entropy measure the total distribution of probable states. Both terms are related to the probability (from 0-100%) that a discrete random variable is exactly equal to some value. That is, both information and entropy are related to the likelihood (or *certainty*) that a system exists in some particular macrostate, with each component (whether a particle, an atom, or a transistor) being in a particular microstate. The more possible component states there are, the more *uncertain* each component microstate, the less likely any given macrostate, and the greater the amount of entropy or information held by the system.

Shannon discussed the nomenclature for his equation with John von Neumann, undecided as to whether he should call his new term 'information' (a term he considered overused, even in 1949) or 'uncertainty' (which seemed to describe the phenomenon well). von Neumann advised him: "You should call it entropy, for two reasons. In the first place your uncertainty function has been used in statistical mechanics under that name, so it already has a name. In the second place, and more important, no one knows what entropy really is, so in a debate you will always have the advantage."

John von Neumann was quite familiar with these equations himself, as he had been working to extend Gibbs' equations of entropy into the field of quantum mechanics [83]. von Neumann found that, if classical statistical mechanics can describe entropy by stating the probability distribution of any set of atoms, then the statistical ensemble of quantum states can be described by a higher-dimensional probability measure called a density matrix. The effect of integrating the additional uncertainty is to increase dimensionality and entropy simultaneously. We will return to von Neumann's work in Chapter 13.

It is apparent from these equations that entropy is a useful measurement of disorder in gases, a useful measurement of information capacity in an integrated digital circuit, and a useful measurement of overall uncertainty in a density matrix of any dimensionality. It is no coincidence that these equations are so similar. Each equation is essentially measuring the same thing: the likelihood of objects being in a certain arrangement. In each case, the sum of all possible microstates for every component in the system yields a probability density for the entire system. And that integrated probability density *is* the quantity of entropy or the quantity of information held by that particle system. The lesson here is that information can be considered a general property of physical systems.

In summary: Information describes the possible arrangements for all the components of a physical system, and any physical system which has many possible arrangements can be used to encode information. That is the case with a network of neurons, each of which can be firing a signal or not at any given time.

CHAPTER 11
How neurons encode information

Neurons encode information – it's what they do. But how exactly do these specialized biological cells accomplish such a complex task?

Just like every other type of cell in the body, neurons perform a particular function. The particular function of neurons is to send and receive electrical signals. *Neurons harness the movement of charged particles to send signals to other neurons.* This is how they encode and transmit information.

Neurons have a specialized structure which helps them achieve this specialized function. These cells are comprised of highly branched dendrites to receive signals, a soma (or cell body) which contains the nucleus, and a highly branched axon to send signals to other neurons (Figure 4). All of our neurons have this general structure, except for sensory neurons, which have an apparatus to convert signals from the external world into electrical signals, rather than dendrites to receive signals from other neurons.

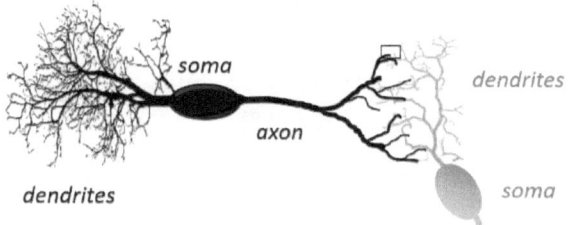

Figure 4. Neurons are specialized cells. They are comprised of highly branched dendrites to receive signals, a soma (or cell body) which contains the nucleus, and a highly branched axon to send signals to other neurons. All neurons have this general structure except for sensory neurons, which have a specialized apparatus to convert signals from the external world into electrical signals, rather than dendrites to receive signals from upstream neurons. Here, one neuron is shown in black and another is shown in grey. The squared region is shown in more detail in Figure 5.

Using special molecule-sized machinery, a neuron will actively pump positively-charged atoms (ions) across its membrane. This process builds up a 'chemical potential' across the cellular membrane, with sodium ions concentrated outside the cell and very few sodium ions inside the cell. This process also causes the cell to build up an 'electrical potential', with the inside of the cell negatively charged compared with the extracellular space. This resting state of a neuron is called the 'resting potential', because the neuron is building up a combined electrochemical potential which can be used to encode information.

Whenever the neuron receives a signal from neurons upstream, some sodium ions trickle into the cell. That tiny flow of current, across the highly resistant neuronal membrane, causes the cell to increase its voltage. When a particular voltage threshold is reached, a large number of voltage-activated sodium channels open, allowing even more positively-charged ions into the cell. The resulting flood of sodium ions is a large flow of current across the neuronal membrane, which quickly and dramatically increases the voltage inside the cell.

And so, when a certain voltage threshold is reached, the neuron is triggered to fire an action potential – an all-or-nothing event where the floodgates embedded in the cellular membrane open, positively-charged sodium ions rush in, and the voltage of the neuron soars. This event is called the action potential. The action potential only lasts a millisecond – one 1/1000th of a second – but it means that the neuron *has sent a signal*. The neuron has briefly flipped from an off-state to an on-state.

The action potential has an effect on small structures within the neuron, called vesicles. Vesicles are made of the same kind of lipid material as the cellular membrane, and they are loaded up with neurotransmitters. During the action potential, the vesicles dock at the inside of the cellular membrane, then fuse with it, releasing their material into the synapse. The neurotransmitters then bind to receptors located on nearby dendrites. Each receptor changes its shape as a neurotransmitter molecule binds to it – opening a channel to permit the entrance of sodium ions into the neuron. That tiny influx of sodium ions causes the post-synaptic cell (the neuron receiving the signal) to increase its voltage. Once again, if the voltage reaches a certain threshold, voltage-activated ion channels will open and sodium ions will flood into the cell, and the entire process will repeat itself (Figure 5).

Every biological neural circuit has these properties. This kind of highly-evolved anatomy and physiology can be found not only in our own brains and spinal cords, but also in the neural circuits of dogs, cats, pigeons, parrots, snakes, fish, and other animals.

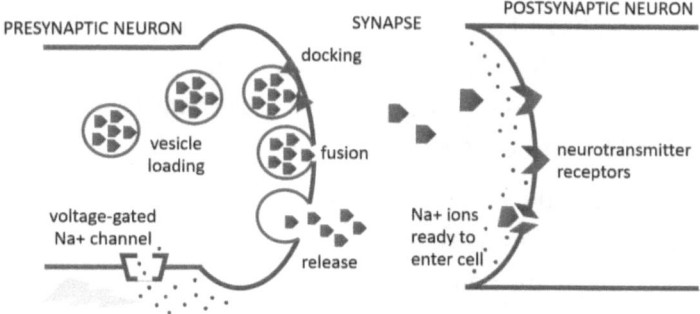

Figure 5. Vesicles loaded with neurotransmitter dock at the cellular membrane, then fuse during the action potential, releasing their material into the synapse. The neurotransmitters bind to receptors on the post-synaptic cell. Those receptors then open channels to permit the entrance of sodium ions (Na+). This influx causes the post-synaptic cell to increase its voltage. If the voltage reaches a certain threshold, voltage-gated ion channels will open and sodium ions will flood into the cell. That event is called the action potential. This sharp increase in voltage causes vesicles loaded with neurotransmitter to fuse with the cellular membrane and release their contents into the synapse, affecting other neurons. In spinal reflex circuits, a single upstream signal may cause a neuron to reach action potential threshold. In cortical neural circuits, the post-synaptic cell must receive multiple input signals within a short period of time in order to trigger an action potential. By integrating multiple coincident upstream signals, cortical neurons calculate whether to send a signal onward.

Yet there are key differences. In spinal reflex circuitry, a single upstream signal can prompt a neuron to reach action potential threshold and send a signal onward. But in cortical circuitry, a post-synaptic cell must receive multiple input signals within a short period of time in order to trigger an action potential. By integrating multiple coincident upstream signals, the cortical neuron calculates whether it will send a signal onward.

We will return to those important differences between spinal reflex circuits and cortical neural circuits in Chapters 14 and 15. But for now, it is worth looking at the characteristics of a generic neuron a little more closely, to think about how information is encoded by the action potential.

As a result of the electrochemical potential across the neuronal membrane, a neuron in a resting state is negatively charged. The voltage inside the neuron, during that resting state, is usually around -70 millivolts (Figure 6). As current flows into the cell, due to the opening of ion channels, the voltage increases. Once a certain threshold is reached – and the exact threshold may be different for every cell – voltage-activated ion channels open and sodium ions flood into the cell. This electrochemical event – called the action potential – is marked by a large and abrupt increase in the voltage across the cell membrane.

A neuron spends most of its time in an off-state, not firing an action potential. It may fire a signal only every few seconds. The neuron is then in an on-state for just a single millisecond – an incredibly short period of time. For this reason, the information encoded by the neuron is not only encoded in the fact that it fires an action potential, but also in the timing of that action potential.

And so there are a few different ways to describe the information carried by a neuron. The state of the neuron can be measured as a continuous function, related to the movement of charged ions, by graphing the voltage changes over time (Figure 7A). Or, the activity of a neuron can be graphed as the number of voltage spikes over a given period of time (Figure 7B). Neurons also fire relative to the entire population of cells; this measurement is often called phase coding, and it can be graphed by comparing single-unit activity to the activity of all neurons (Figure 7C).

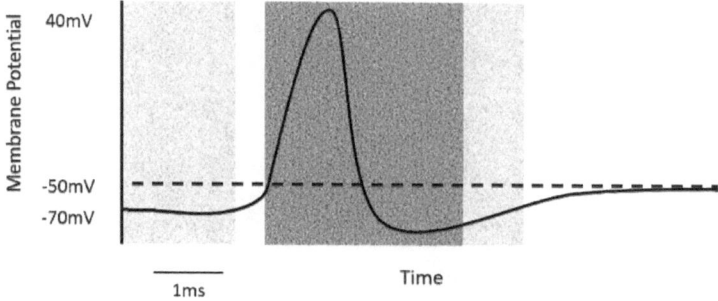

Figure 6. A neuron harnesses the movement of charged particles to fire an action potential, which triggers the sudden release of neurotransmitters into the synapse. The grey region of the graph shows the cell at resting potential, approximately -70 millivolts (mV). In the white region of the graph, the cell is highly sensitive to ion flux, as it reaches the voltage threshold for firing an action potential. In the dark grey region of the graph, the cell is firing an action potential. In this moment, voltage-gated ion channels open and sodium ions (Na+) rush into the cell. Immediately afterwards, potassium ions (K+) flood out of the cell to reinstate the electrochemical difference. The cell then returns to its resting potential, able to fire an action potential yet again.

Neurons that are sparsely distributed across the brain will often fire together. This is known as sparse coding. Sparse coding is characterized by the number of active neurons in the population at a given moment in time. While each neuron has a distribution of responses (e.g. over some range of sensory inputs), the responses of many sparsely-distributed neurons, in combination, encode the external state of reality (e.g. describing the 'what' and 'where' details for a particular sensory stimulus). Temporal coding, rate coding, and phase coding are separate methods to analyze the amount and the content of data that is carried by individual neurons, by themselves and in relation to each other.

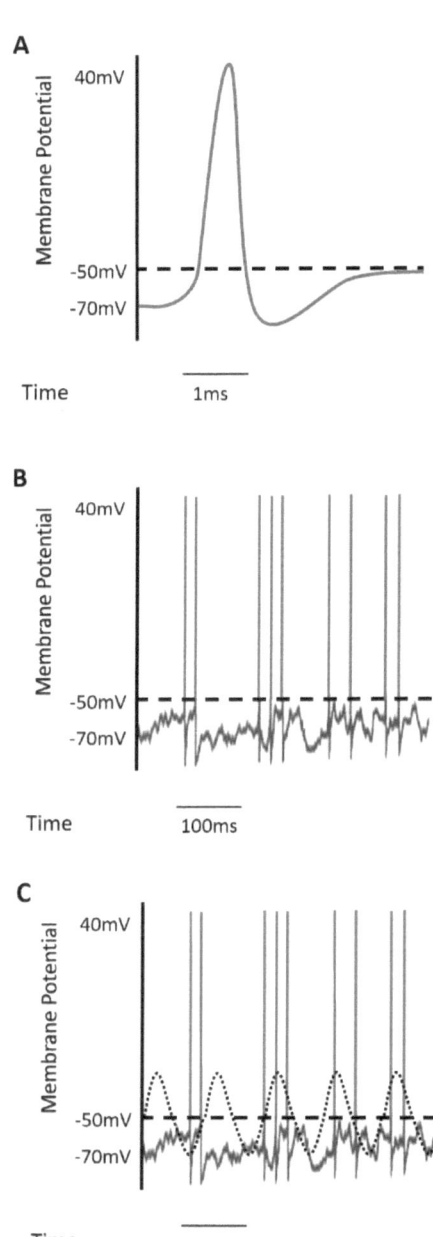

Figure 7. Upon stimulation, a neuron that was at resting potential (approximately -70 mV) may or may not reach the threshold for firing an action potential (approximately -50mV). If the threshold voltage is reached, voltage-sensitive ion channels open and positively-charged ions flood into the cell, causing a massive depolarization (reaching about +40mV). Then the cell repolarizes, by pumping positively-charged ions out of the cell. This process results in a transient hyperpolarization or refractory period, during which time the cell cannot fire an action potential. The cell then returns to resting potential; the electrochemical gradient is restored and positively-charged sodium ions are again concentrated outside the neuron, once again ready to flood into the cell when channels open. **(A)** This graph shows a cell reaching threshold and firing an action potential, undergoing hyperpolarization, then receiving further stimulus that does not quite reach threshold. The voltage measurement of the neuron provides a constant analog signal. The *timing* of the action potential is subject to probabilistic factors at the level of electrons, and the action potential itself is a low-probability event that encodes information. **(B)** This graph represents the same cell, but over a longer period of time. The state of being in the midst of firing an action potential or not is binary; in this way, neurons are capable of producing an effectively digital signal. The *rate* at which a neuron fires action potentials thereby encodes information. **(C)** This graph demonstrates how the activity of a single neuron (solid line) takes place 'in phase' with the activity of the entire neuronal population (dotted line). This is called phase coding, as the action potential of an individual neuron occurs relative to the *phase of the oscillation*. This relationship also encodes information.

In short, neural activity is coordinated across sparsely-distributed neuronal populations through highly synchronous activity. That is, many neurons across far-flung regions of the brain will fire together, in phase with each other [84]. When this synchronous neural firing occurs at regular intervals, this activity is called a neural oscillation. Phase coding relates individual neuron firing rates to the oscillatory activity of the entire cell population.

These regular bursts of synchronized neural activity, spreading across broad regions of the brain several times per second, occur at a range of frequencies (Figure 8). Synchronous activity has been observed at the following frequency bands: theta (4-8 Hz), alpha (8-12 Hz), beta (13-30 Hz), and gamma (which includes a wide range of high-frequency oscillations, from 30 to 150 Hz). Neurophysiological studies have demonstrated the presence of strong, synchronous global neural activity during sleep, coma, and anesthesia, along with alterations in the frequency spectrum [84]. Higher-frequency oscillations are often embedded within the lower-frequency oscillations, with several of the high-frequency events taking place during a single period of a slow-frequency oscillation.

Neural oscillations are the result of the coordinated firing of sparsely-distributed neurons. It is worth noting that oscillations are not observed in spinal circuitry, only in brain circuitry. For many reasons, this periodically-occurring synchronous activity is thought to be critical for the binding of perceptual experience, memory retrieval, and optimization of behavioral output [85,86].

Slower oscillations, particularly theta rhythm, originate from hippocampus and increase in power during the sensorimotor integration phase of task learning [87]. Meanwhile, gamma frequency oscillations originate from the midbrain but manifest across the thalamus and neocortex, enhancing the processing of sensory information by routing gaze or auditory attention toward the external stimulus [88].

As a result of these and many other studies, oscillations are thought to solve the binding problem, joining different brain regions in the time dimension by enhancing synchronous neural activity. Introducing gamma oscillations in experimental settings

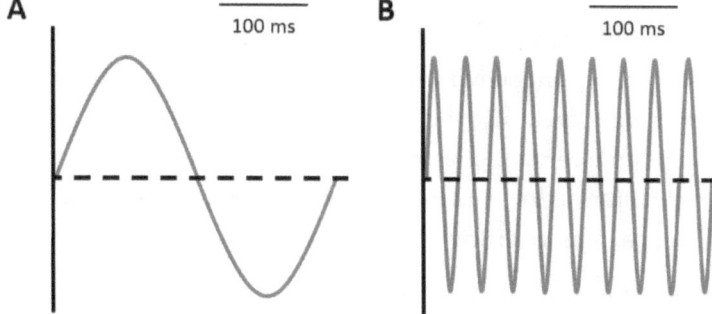

Figure 8. Synchronous activity has been observed at a range of frequency bands in cortical neural networks, including: **(A)** theta (slower oscillations, with a frequency of 4-8 Hz) and **(B)** gamma (faster oscillations, with a frequency of 30-150 Hz).

even enhances perceptual binding across sensory modalities in a working memory task, in a manner dependent on the phase of the induced stimulus and the person's own gamma oscillation [89].

Intriguingly, low-frequency oscillations have been shown to play a critical role in the consolidation of memory. Specifically, the action potential induces calcium release within the cell body, and that stabilizes synaptic connections by promoting pro-neural gene transcription and synaptic remodeling [90]. Higher-frequency oscillations also contribute to remodeling the anatomical and physiological features of participating neurons. For example, synchronous activity occurring at fast gamma frequencies has been shown to optimize firing patterns of individual cells and prompt the activity-dependent survival of neurons which have been recruited to that synchronous activity [91].

The coordinated firing behavior of sparsely distributed cortical neurons is shaped over development and remodeling across the lifetime. *But what leads a system to take on this ordered state*, with coordinated firing patterns that encode events happening in the surrounding environment? The answer to this question is frustratingly cyclical: synchronous or sequential neural activity prompts the remodeling of synapses, so these synapses better encode information, and that remodeling process allows neurons to engage in more efficient synchronous or sequential activity.

The spontaneous remodeling of neural connections is absolutely critical to the function of complex neural circuitry. The brain actually changes its physical structure to store memories!

The biological rule guiding this remodeling process is called Hebbian plasticity. In neural networks, 'cells that fire together wire together' [92,93]. Previous experiences, encoded in ongoing neural activity, are actually stored in the structural connections between neurons. That way, a similar event in a similar context can provoke the same sequential pattern of neural activity, so that we can handle a familiar situation with ease.

In summary, neuroscience is able to explain quite a lot about how neurons process information. We understand how neurons transmit information, by setting up a resting potential and then firing an action potential. We understand how neurons encode information in the timing of the action potential, the firing rate, and even the relationship between individual signaling events and the synchronous activity of the neuronal population.

And finally, we understand how neurons store information, by remodeling their synaptic connections to favor sequential firing patterns that have proven useful in the past. However, there are some key features of cortical neural networks that are still not well-explained within classical neuroscience.

CHAPTER 12
What neuroscience cannot explain

Neuroscience today can explain a lot about how information is encoded, transmitted, and stored in the brain. But right now, the field cannot explain everything.

We understand exactly how photons of light hitting the retina are converted into electrical signals, then sent to processing centers in the brain – where neurons in the visual cortex encode different aspects of the visual field. We also understand how sound waves hitting the ear are converted into electrical signals, then sent to processing centers in the brain – where neurons in the auditory cortex encode various aspects of the sound, like volume and pitch. In short, we already understand very well how information is *encoded* in the brain.

But we do not understand how the encoding process gives rise to this cohesive stream of multi-sensory perceptual experience. Why should there be *qualitative content* associated with cortical information processing, when that is simply not the case with spinal reflex circuits – or, seemingly, with any classical computing architecture? The connection between cortical neural activity and *qualia* remains unexplained.

Secondly, we understand well how information is *transmitted* in the brain. We know that neurons send signals to each other, by setting up the resting potential and then, if upstream signals converge, firing an action potential. We understand how cortical regions that are dedicated to integrating information across sensory modalities compile a messy deluge of upstream signals into a cohesive output signal, which in turn guides the action of the body. As a result, we understand how the synchronous timing of sparsely-distributed neurons across the neocortex correlates with the initiation of voluntary behavior.

But we do not understand how non-deterministic computation in cortical neural networks connects to the exertion of free will. Why do all other systems obey the laws of classical mechanics, while human beings and other living animals spontaneously get up and move around, deliberately choosing our actions in the world to achieve our goals? This purported link between cortical neural activity and *causation* remains unexplained.

Thirdly, we understand how information is *stored* in the brain. We understand how sequences of neural firing, which reliably occur in response to a sequential pattern of events in the local environment, trigger the remodeling of synaptic connections between cortical neurons. We have discovered many of the biological mechanisms underpinning this synaptic remodeling process, and we know this synaptic remodeling process favors those same sequential patterns of neural activity to re-occur, particularly in similar contexts. It is useful to reliably engage familiar patterns of thought and behavior in a familiar context.

However, we simply do not know why human beings and other living animals spontaneously grow into a more ordered state over time. That capability seems to disobey the second law of thermodynamics, which states that natural systems tend to get more *disordered* over time. The mechanism connecting cortical neural activity and *spontaneous self-remodeling of the system into a more ordered state* remains unexplained.

The field of neuroscience is currently able to explain well how incoming sensory data is encoded, transmitted, and stored by neurons, and how that leads to motor output (Figure 9). This model of neural function provides a very good description of spinal reflex circuits, accounting for all observed phenomena. Since there is no streaming perceptual experience, no memories or cognitive models, and no voluntary, goal-directed movement emerging from these simple circuits, all features are well accounted for, under classical assumptions.

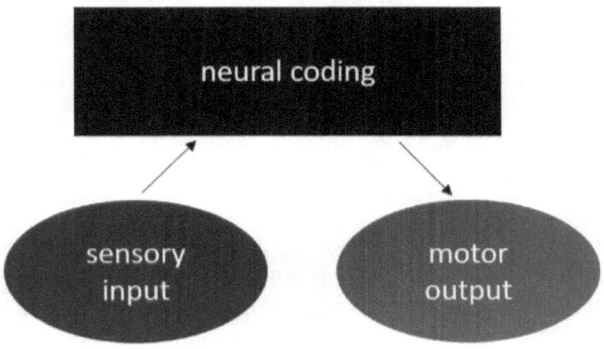

Figure 9. Classical neuroscience is able to explain well how sensory input is encoded, transmitted, and stored by neurons. This model of neural function provides a very good description of spinal reflex circuits.

However, this model of neural function does not provide a very good description of all observed phenomena emerging from cortical neural network activity. Certain features are not fully accounted for, specifically: perceptual experience, spontaneous self-remodeling into a more ordered state, and voluntary action.

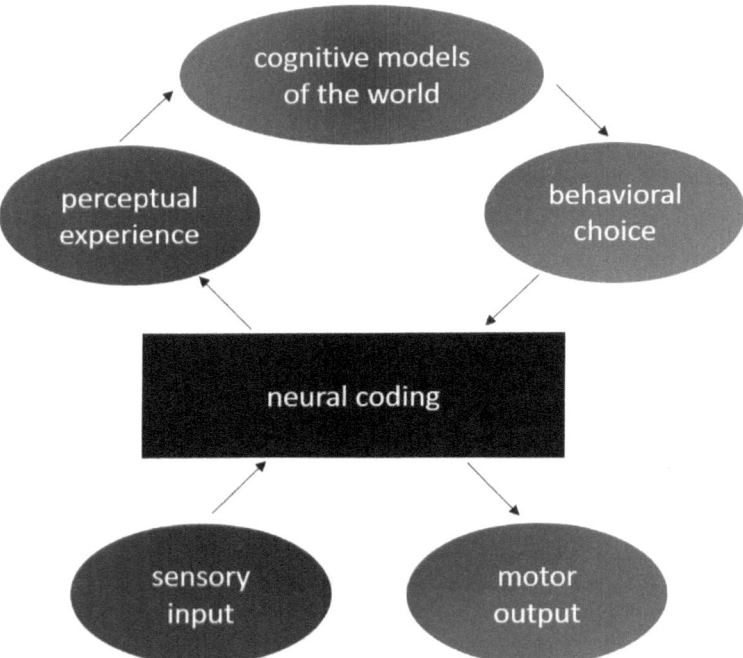

Figure 10. Classical neuroscience is *not* able to explain how neural activity gives rise to 1) streaming perceptual experience; 2) cognitive models of the world, centered on a concept of the self, which grow over time, as the system stores memories and remodels itself into a more ordered state; and 3) spontaneous actions, which achieve goals to protect the existence of the self. The current model of neural function simply cannot account for all phenomena emerging from cortical neural network activity.

In short, the field of neuroscience is very good at explaining how simple neural circuits work. But something goes wrong when we try to apply everything we know about spinal reflex circuits to cortical neural circuits. We're missing something critical here. All of the key features of consciousness seem to emerge from cortical neural activity, but not from spinal reflex circuitry.

Let's take an example. If a person touches a hot stove, they will remove their hand immediately, before they have any conscious awareness of *having that pain* or *selecting that action*. This neat and efficient reflex is achieved by a simple neural circuit called a spinal reflex circuit.

This simple reflex circuit is composed of a sensory neuron, an interneuron, and a motor neuron. This useful type of three-neuron circuit is ubiquitous in our spinal cords. When we touch that hot stove, a little temperature receptor embedded in the skin is activated. That temperature receptor is located on one end of a sensory neuron. The sensory neuron receives this signal from the external environment and converts it into an electrical signal – a flow of charge into the sensory neuron, indicating that something has happened. That sensory neuron then sends a signal all the way to an interneuron, located in the spinal cord. That interneuron sends a signal directly to a motor neuron, which activates flexor muscles and inhibits extensor muscles, achieving an effective limb withdrawal reflex. At the same time, that sensory neuron sends a signal to *another* interneuron, which crosses the spinal cord and signals to another motor neuron on the other side of the body. That motor neuron activates extensor muscles, and inhibits flexor muscles, to steady the contralateral limb. This spinal reflex circuit achieves a perfect limb withdrawal response, without any actual conscious experience of pain.

It's only later, as ascending neural circuits deliver that important sensory information to the brain, that that person will have any *conscious experience* of the pain. That qualitative perception requires the activation of neurons in regions of neocortex that are dedicated to processing the incoming sensory information.

If there is no activation of the relevant cortical neural circuitry, then there will be no conscious perception of pain.

So there must be something unique about cortical neural circuits – a way of encoding information in a fundamentally different way than spinal reflex circuits, which gives rise to qualitative information content.

This is not just about having greater computational power. This is about differentiating between computational processes that give rise to qualitative content and computational processes that do *not* give rise to qualitative information content. So let's take some time to consider how cortical neurons encode information, and how that encoding process is different from the encoding process in spinal reflex circuits (Figure 11).

Cortical neurons are not activated by a single external stimulus event, like sensory neurons, or a single upstream signal, like spinal interneurons and motor neurons. Cortical neurons have the unique feature of *integrating* a whole lot of upstream signals over a short window of time – and that may or may not reach the threshold for sending an output signal. What's particularly interesting about this *integrative process* is that it is a process of noisy coincidence detection – cortical neurons allow stochastic charge flux, and random fluctuations in membrane resistance, to affect the probability of a state change. And so, it is incredibly hard to predict the firing patterns of individual cortical neurons,

Spinal Reflex Circuits	Cortical Neural Circuits
neural activity **is not** associated with streaming perceptual experience	neural activity **is** associated with streaming perceptual experience
simple coding a single suprathreshold stimulus will easily trigger an action potential	**complex coding** multiple coincident events are needed to trigger an action potential
classical outcomes spontaneous fluctuations in membrane potential *are not* observed	**probabilistic outcomes** spontaneous fluctuations in membrane potential *are* observed
clean signals firing outcomes are robust to random electrical noise	**noisy signals** firing outcomes are sensitive to random electrical noise
deterministic signaling outcomes *can be* predicted by summing upstream inputs	**non-deterministic** signaling outcomes *cannot be* predicted by summing upstream inputs

Figure 11. Some key differences between spinal reflex circuits and cortical neural circuits are listed here. In cortical neurons, multiple upstream signals are integrated with random electrical noise to affect the probability of sending a signal.

which are dependent on all kinds of upstream factors and essentially random noise. Meanwhile, it is incredibly easy to predict the output of neurons in spinal reflex circuits – these signaling outcomes are completely deterministic, while cortical neuron signaling outcomes are *probabilistic*.

The unpredictable behavior of cortical neurons is not currently well-explained in neuroscience. Researchers tend to use statistical methods to describe spiking patterns at the level of entire cell populations. However, it is very difficult to pinpoint the timing of individual cortical neuron firing, the way it is possible to pinpoint when, where, and why a spinal interneuron will fire. We just do not understand the physical mechanisms underlying spontaneous state change in cortical neurons.

What is particularly difficult to understand is why these neurons would retain sensitivity to stochastic charge flux, when lower neural circuits have evolved to be so robust to random electrical noise. Cortical neural networks even organize into coordinated states, called up-states, where nearly every neuron in the cerebral cortex is sensitive to stochastic charge flux for a short temporal window. Why would cortical neurons work so hard to make themselves sensitive to random electrical noise?

Random electrical noise reduces accuracy and energy efficiency in classical circuits. It is wasteful to have all this uncertainty, disorder, and randomness in the system, when calculating binary outcomes.

So why would cortical neurons be so noisy? And how could they possibly be so accurate and energy-efficient, despite retaining a high sensitivity to random noise? Perhaps, our highly evolved cortical neural networks have not established some new way to be more inefficient, but rather are using that noise to drive a categorically different, *non-deterministic* form of computation.

CHAPTER 13
Information: probability and uncertainty

To really dig into the physical relationship between information, probability, uncertainty, and the non-deterministic nature of our world (and ourselves), we must return to the foundational work of John von Neumann, who is generally considered one of the greatest minds of the twentieth century. John and his wife Klara fled their home country Hungary during World War II, settling in the United States. Initially, they worked at Princeton – he as a professor at the Institute for Advanced Study, and she as a statistical analyst at the Office of Population Research. Later, the couple spent time at Los Alamos Laboratories, where they built and programmed some of the first computers. John wrote on a wide range of topics in mathematics, information theory, and quantum mechanics. After the war, he started writing about the similarities between the computer and the human brain. Before he could finish, he tragically died of cancer. Klara edited and published his lectures, then walked into the Pacific Ocean.

It turns out that physical systems, including people, do not act deterministically. John and Klara von Neumann taught us that, in a million different ways. They believed that human beings are subject to the laws of nature and *are capable of computation*. To push these ideas forward, it is worth thinking deeply about the relevant laws of physics, in order to understand how information-processing systems really work. The relevant laws of physics are the laws of quantum mechanics.

Quantum mechanics may seem to contradict the laws of classical physics, with baffling implications for the nature of reality. But it is worth digging into how these processes really work. While it may be easier and more comforting to rely on classical physics, which expects matter to act deterministically, it is useful to admit that matter, at the most fundamental level, acts probabilistically. This may initially seem hard to reconcile with our experience of reality, yet the truth is that our world actually does consist of probabilities, rather than certainties, at the quantum level. Ignoring this difficult fact is not worthwhile in the long run. Accepting this difficult fact offers us the opportunity to build a greater understanding of reality and to advance our technology.

Let's return again to the past. In the 1920s, the physicist Werner Heisenberg proved we could not know both the exact position and the momentum of a particle at the same time. This is not due to a limitation of our measuring systems; it is a fundamental law of nature. Particles essentially exist in the present moment as a likelihood of being in one place or another, moving in one way or another, rather than in a defined state [94]. There is a fundamental indeterminism here, which renders the particle more of a *probability* than an actual, physical thing. This is worth repeating: a particle is indeterminate – *probabilistic* – in the present moment.

The likely position and momentum of a particle – in relation to space and time – was later represented in an equation, derived by Erwin Schrödinger [95]. This equation describes the 'eigenstate', or state of quantum superposition, for the particle. Because the particle is not defined, but probabilistic, the particle itself is best described as a set of fluctuating probability amplitudes, known as a wavefunction. The Schrödinger equation is a mathematical description of a particle's possible states, or its wavefunction – and critically, it also accounts for the *constraints* that particle exerts on other particles in the same system. Two particles just cannot occupy the same position, momentum, and atomic orbital. If two particles did end up having these exact same properties, then they *would be* the same particle, and the universe would lose mass. Therefore, this impossibility cannot occur – a fact known as the Pauli exclusion principle, after a contemporary of Heisenberg and Schrödinger named Wolfgang Pauli.

The idea of probabilistic states forms the foundation of quantum mechanics. It is counter-intuitive to our experience, because all the constraints particles place on each other lead probabilities to reach certainties at the scales we normally observe. Only if we peer closely at the quantum realm do we find that uncertainty.

This uncertainty can be quantified, in the form of the Heisenberg uncertainty principle. This mathematical law provides that: the range of possible *positions* of a particle, multiplied by the range of possible *momentums*, is always greater than some minimum value. These values are probabilistic, not deterministic.

Specifically, the range of uncertainty in the *position* of an object and the range of uncertainty in the *momentum* of that object will always have a value that is greater than the Planck constant h divided by 4π.

Any particle – a photon, an electron, even a whole atom – can be described as a set of possible states which describe the most likely trajectory of the object across space and time. Yet the likely position and momentum of the particle depend on the actions of other particles, which are also probabilistic. As a result, the entire system must be considered as a whole – as a single, more complex wavefunction. Because position, mass, energy, and time are always multiples of discrete units (called Planck units), the system state can either be described as a density matrix or as a wavefunction – either mathematical representation provides all possible values for each particle. While Schrödinger preferred the continuous wavefunction method of describing possibility, Heisenberg preferred to use the matrix mathematics method, which seemed to him truer to the discrete nature of particles.

Uncertainty does not just apply to position and momentum. It also applies to other properties of particles, such as the amount of mass-energy they have, and how much time has elapsed since that value was last measured [96,97]. Time, in the quantum world, is simply not defined – except in relation to the total mass-energy of an object that exists in time. The more certain one is about the time elapsed since the last measurement, the less certain one can be about the total mass and energy of that object.

We can do a little thought experiment here, to explore this notion. We can imagine a particle, traveling alone, in a pure vacuum, across the universe. Without undergoing any interactions, its mass and energy will remain the same. But without collision or detection of any kind, it simply cannot be said how much time has passed. Conversely, the uncertainty in the energy of the system *must increase* whenever the uncertainty of the time taken to measure the system *decreases*. As a result, any particle system that is capable of measuring time exactly – for example, by employing some system-wide process of coincidence detection – may be capable of expanding the range of possibilities for how all of the matter and energy in the system is arranged. The mechanism for such an astonishing feat is simply applying the Pauli exclusion principle across the entire particle system.

Another physicist, Niels Bohr, recognized this close relationship between mass-energy and time, and so he derived an equation to describe this uncertainty relation. Incredibly, his equation gave the exact same result as Heisenberg's uncertainty principle. The range of uncertainty in the energy of an object multiplied by the range of uncertainty in the time taken to measure that object will always have a value greater than the Planck constant h divided by 4π. That is: time/energy uncertainty has exactly the same minimum value as the position/momentum uncertainty. There are clearly fundamental limits at work here.

Again, this uncertainty is not caused by some limitation of our current technology; it is a fundamental law of our universe. The standard deviation of the energy held by a particle – a statistical quantity – may decrease only when the standard deviation of the time taken to measure the system *increases*.

Since the speed of light is always constant – with a massless object always traveling at 299,792,458 meters per second – the inability to say how much time has elapsed since a particle was last detected renders problems for discovering where exactly in space the particle might be located after that uncertain period of time.

Indeed, there is one more fundamental uncertainty principle in quantum mechanics. This is the relation between energy and position. One cannot know both the amount of mass-energy held by a particle *and* its exact position. This uncertainty principle is marked by a familiar minimum value: once again, the range of uncertainty in the energy of an object and the range of uncertainty in the spatial position of that object will always have a value greater than the Planck constant h divided by 4π.

Particles only have some 'real' value once they are detected by another particle or particle system, by measurement or collision. That is not normally how we think about the world. Normally, we expect that objects have stable characteristics, regardless of how they interact with other objects. Yet that is not the case at the particle level. The properties of electrons and photons are really only defined by their interactions with other particles. If there are no interactions, each particle remains in an undefined state. The particle exists in a density of possible states, rather than a single defined state.

So, how can the values for position, momentum, and energy be calculated, in practical terms? The likely values can be estimated statistically. This was the approach taken by Max Born, another contemporary of Heisenberg, Schrödinger, Pauli and Bohr.

Born realized that, after some uncertain period of time, the possible positions and momenta for a particle were described by a probability distribution – given by Schrödinger's wavefunction or Heisenberg's matrices. As a result, there was some statistical likelihood the *detected* position and momentum for a particle would be a particular value from that set.

Exact solutions to these wavefunctions or matrices have also been found; these solutions correspond to the abrupt collapse of alternative eigenstates as the exact measurement of a particle is made and all other probable states disappear from existence.

The process by which this happens is a foundational principle of quantum information theory. And that is why we must return to the work of John von Neumann. He was the one who realized the uncertainty principle meant that our universe was inherently probabilistic, that probabilistic particle behavior naturally created a set of possible system states, and that quantity was essentially *entropy* or *information*. John von Neumann was the person who connected the concepts of quantum mechanics to the existing fields of thermodynamics and information theory.

In order to explain how probabilistic particle behavior gave rise to information entropy, von Neumann introduced the notation used in quantum mechanics. He described particles not as real objects, but as probabilities, represented mathematically with an algebraic shorthand called a density matrix. By multiplying two density matrices together, he could describe what happened whenever two particle systems interacted with each other.

As two particle systems interact, their probabilities interact with each other. This is the same as saying their wavefunctions or their matrices interact with each other! Either description is valid. The state of every particle in the system must be compatible with every other particle – they cannot exist in the same position, with the same properties, so these outcomes are just excluded. Introducing these constraints, as particles interact with each other, *collapses the wavefunction* or *compresses the density matrix* because fewer compatible states are now possible.

The interaction of two particle systems reduces the total number of possible system states, as redundancies or consistencies are identified. As the systems find the most compatible arrangement, alternative probabilities fall away and the wavefunction collapses. The total information held between the two systems is compressed.

As these solutions were being found, another great mind began to tackle what happened when the wavefunction collapsed, the density matrix resolved, and information became compressed. That great mind belonged to the physicist Richard Feynman. While exploring the mechanics underlying this process, Feynman discovered that any resolution of a wavefunction translates to discrete changes in the behavior of the participating particles [98].

Upon the detection, observation, measurement, or collision of particle systems, the position, momentum, and atomic orbital of every electron in the system is defined, and the interactions between atoms (the forces they exert on each other) can then be deciphered using the equations of classical electrostatics – that is, as long as one accounts for the charge distribution across each atom. But critically, the resolution of the wavefunction is accompanied by a slight change in the distribution of electrons around each atom in the system, resulting in a small shift in the angular momentum of each of the atoms in question.

In other words, any detection or collision event that leads to resolution of a particle state *actually changes reality*, as the charge distribution of each atom across the now-defined system is shifted. This finding, known as the Hellman-Feynman theorem, precisely reconciles the fundamental uncertainty principles of quantum mechanics with the laws of electromagnetism in thermodynamic systems. This theorem permits us to describe *how quantum oscillations affect our reality*.

The Hellman-Feynman theorem says: the change in momentum for each electron underlying the wavefunction is proportional to the change in any parameter that contributes to the total energy of the system, such as a shift in the local electrical field or a shift in the position of other electrons in the vicinity.

Therefore, any perturbation *to the system* can cause a change in the momentum *of any electron* in the thermally-bound system, in a manner related to the original unperturbed states and the volume of possible states. What Feynman found is that once particles interacted with each other, their possible states become constrained. These constraints allow the position of each electron to be defined in a compatible way with every other electron in the system, in accordance with Pauli's exclusion principle.

But this event is associated with an alteration in the organization of electrons around the nucleus, with the charge distribution distorted from central symmetry. That causes a dipole moment, or a small boost to the angular momentum of each electron. The smaller the newly-defined distance between atomic nuclei, the larger the dipole moment experienced by their electrons. In this way, neighboring atoms can exert effects on others' behavior [98]. This shift in charge distribution causes each atom to instantly change its angular momentum, and that effectively restores the uncertainty in position and momentum.

As a result, a defined system state never really exists in the present moment. Upon collision or detection, the quantum state is resolved into a singular reality – but then it all starts changing again. This is called the von Neumann projection postulate. That is – for any measurement with discrete results, which does not destroy the entanglement of the system, every measurement prepares a *new* state. Some uncertain amount of time will pass, and the system will generate a new probability density [96].

It is worth noting that all particles within a closed system share a single wavefunction or density matrix. Resolution of any one component therefore leads to resolution of all components, in accordance with the Pauli exclusion principle.

It is worth pondering what quantum information, probabilities, density matrices, or wavefunctions *are*. We might consider the physical properties of these curious mathematical constructs: According to the superposition principle of quantum mechanics, individual wavefunctions can be added together and multiplied by complex numbers to form a Hilbert space, or a complex plane, which is populated by the probability amplitudes of all particles in the system. These probabilities are considered to exist only in that abstract mathematical space – not within our observable universe, but rather defined along a 'complex plane' which is perpendicular to all of the observable dimensions of space and time. Particles really only exist in that finite-dimensional vector space. Yet it is not clear whether this extra-dimensional space – or the probabilities themselves – are an actual part of our reality, or just a convenient mathematical abstraction [99,100].

Again, wavefunctions or density matrices are just mathematical descriptions of the quantum state of a particle system. They represent all probabilities, defined along a complex plane. Each method describes all of the possible microstates for each particle in the system, at the point of measurement. As such, they describe the *information* contained in the system.

The total number of possible microstates in a physical system is, mathematically speaking, the amount of entropy or information contained in the physical system. It is the sum of all uncertainty, the sum of all probabilistic states of every particle in the system. And so, the quantum nature of particles is deeply connected to the amount of information they can hold. The process of *creating a density of possible system states*, then *reducing that density of possible system states by interacting with another particle system*, is a natural process of quantum computation.

Let's recap. The position and momentum of every particle in a system are uncertain, and dependent on each others' actions. This quantum state can be described mathematically as a wavefunction or density matrix comprising all possible states. Each atom in the combined system must take a distinctive stance in relation to its momentum, position, and energy state. After all, particles cannot be identical – this impossibility is prevented by the Pauli exclusion principle. The quantum state is resolved as two particle systems interact – and that event is associated with a shift in the momentum of every component atom, in accordance with the Hellman-Feynman theorem. Due to those subtle changes in charge distribution caused by the interaction itself, each atom is pulled in the direction of its neighbors, to varying degrees.

A system-defining event is therefore immediately followed by the formation of a new probabilistic state, which evolves over time, in accordance with the von Neumann projection postulate. And as time moves forward, the interactions between systems will continue to occur. Probabilities will continue to propagate and then resolve. By following the laws of quantum mechanics, we uncover a natural – cyclical – process of computation.

There are two steps to the process: information generation and information compression, which occurs as *meaning* is extracted.

The first step involves *creating* some quantity of information (increasing the number of possibilities), as the system permits a random, disorderly movement of charged particles. It's worth noting that any thermodynamic system creates both entropy (in a physical sense) and information (in an abstract mathematical sense). These two values are operationally equivalent, as they are both proportional to the amount of uncertainty in the system.

The second stage involves *reducing* that mathematical quantity of information by allowing constraints to define the *most likely* system state. This is a process of extracting *consistencies* from that information. As constraints are introduced, the enormous range of possible system states grows more limited and the most likely system state emerges. The wavefunction collapses, the density matrix is reduced, and the quantity of information is compressed. This event proceeds in accordance with laws of mechanics. All those possible system states – all that entropy – falls away as a single system state is actualized. That state immediately becomes the past as a new set of possible system states emerges.

It is possible for information to be generated and compressed in a cyclical manner, as particles interact with each other and probability densities are reduced to single actualized states. But not every thermodynamic system can do this! Most systems cannot sustain uncertainty for long enough to complete any computation, and all of the more unlikely possibilities are lost. These systems will act in accordance with classical mechanics, never doing anything spontaneous or unlikely. A good example of this kind of system is the steam engine – it creates a lot of entropy, but cannot select an unlikely, ordered, *meaningful* state.

The key property of the steam engine and other deterministic systems is that they release heat into the environment, rather than acting as a net heat sink. That prevents thermal energy from being effectively held together in a coherent, probabilistic state. And that prevents quantum computations from occurring! Quantum processes only contribute to the behavior of a macro-scale system if probabilities are sustained for timescales that last longer than heat dissipation dynamics. If thermal energy is dissipated too quickly, then uncertainty cannot be sustained.

In this way, macro-scale systems lose their probabilistic nature and act in a deterministic way. The probabilities are still there; they just cannot affect any outcomes. For this reason, most systems which exist at the scale we can observe do not appear to be probabilistic, but rather classical and deterministic.

It is certainly worthwhile to investigate how and under what conditions these processes work. Progress in understanding our physical reality will help us understand the relationship between thermodynamics, consciousness and computation.

In this chapter, I described how the physical characteristics of a particle are intrinsically uncertain, how this uncertainty is related to probability, how probabilistic particle states can be described mathematically, as either wavefunctions or matrices, and how these probabilities can be resolved. In the following chapters, I will explain how biological neural networks might follow these physical laws to cyclically generate and compress information.

Our goal here is to relate events in the nervous system with equations describing entropy, information, and the probabilistic behavior of particle systems. This requires a consideration of neuroscience from the perspective of quantum physics. That does not mean embracing quantum mysticism! It means striving to understand the laws of thermodynamics and mechanics – and how they achieve a physical method of information processing.

In the following chapters, I will demonstrate how incorporating the laws of thermodynamics and mechanics into neuroscience provides an explanation for qualitative perceptual experience and non-deterministic behavioral outcomes. To build this new approach, let's consider how neurons harness the intrinsically probabilistic movement of electrons to encode information.

CHAPTER 14
A mechanism for bottom-up processing: cortical neurons physically create information

The key to discovering a mechanism for the phenomenon of consciousness lies in thinking about neural activity as encoding *information* in a physical manner.

Information *increases* as the probability of any one outcome or system macrostate *decreases*. Imagine the number of possible macrostates that a neural network with 86 billion neurons could take! The number of possible combinations here is large, so there is a large quantity of information.

But a modern computer chip might have 86 billion transistors – what makes the brain any different from such a machine? Why do brains *represent* the information they're encoding, with a stream of perceptual experience that is constantly updated with newly-arriving sensory data? Why are brains capable of general problem-solving abilities and dexterous movement? What makes our computational power and our behavior so much more complex than regular old computers?

We can think about neurons as encoding a digital signal, which corresponds to the cell spiking or not at a given moment in time. But that signal only proceeds if the noisy analog signals percolating through a neuron reach a certain voltage threshold and trigger an action potential. That is not true in transistors, or in simpler neural circuits like those found in the spinal cord. In *cortical neurons specifically*, the random movement of electrons contributes to the voltage of the cellular membrane. As a result, the electrochemical potential, or voltage, of that neural membrane spontaneously fluctuates. Sometimes these spontaneous events are sufficient to boost the neuron over the threshold for sending a signal [101-103]. That is simply not true in spinal reflex circuits and the neural circuits of insects. These have essentially deterministic signaling outcomes, which can be easily predicted by summing all upstream signals impinging on that neuron [104-106].

The inherently probabilistic nature of cortical neuron signaling outcomes is absolutely key here. Probabilistic particle behavior actually contributes to signaling outcomes in the brain, with the resulting computations leading to non-deterministic behavior. And it must be noted that the best way to describe inherently probabilistic events is with quantum mechanics.

To consider how the brain engages in computation, we need to return to the relationship between entropy and information. The movement of charged particles in a neural network creates *both entropy and information* – both of these concepts are *directly* related to the amount of randomness in the system and *inversely* related to the probability of observing any one arrangement of component particle states. Critically, cortical neurons harness the probabilistic movement of charged particles in the vicinity, creating *entropy* – and, in doing so, they convert external events into electrical signals which carry *information*. Let's explore how cortical neurons accomplish this task.

It is often said the firing of an action potential "causes" another cell to fire an action potential, but this is not strictly accurate. In the cerebral cortex, signals received from upstream neurons only slightly increase or decrease the probability of an action potential occurring. To reach threshold, multiple simultaneous inputs must occur, causing multiple ion channels to open at the same time, triggering coincident charge flux. In addition, random ion leak across the neural membrane can contribute to the cell reaching voltage threshold and firing an action potential [107].

As we might recall from the previous chapter, the Heisenberg uncertainty principle states that the position and momentum of a particle cannot both be known [94]. This uncertainty provides a valuable opportunity for information generation: Since every electron within a cortical neural network exists in a probabilistic state, and the voltage state of each cortical neuron is dependent on the exact position and momentum of all these electrons, each neuron exists in a probabilistic state.

Although cells do organize charged particles, pumping them across the membrane, individual particles still exhibit random motion across the neural membrane. This random movement, a function of both particle position and particle momentum, in turn affects the electrical resistance of the neural membrane. Because the position and momentum of any component electron cannot be known, the physical interactions between ions and other ions (or ions and ion channels embedded in the neural membrane) cannot fully be determined. The entire neuron therefore exists in a probabilistic state, sensitive to quantum-level noise.

The random motion of each electron will contribute to whether changes in voltage will sum to action potential threshold within a given window of time, for a given region of neural membrane. If the threshold is reached, voltage-sensitive ion channels open and the cell is flooded with positively-charged ions, causing the enormous voltage spike that is characteristic of the action potential. If the cell does not reach voltage threshold, it returns to resting potential and waits for coincident signals that do boost it to threshold. To predict whether that cortical neuron reaches the threshold for an action potential, we need to describe all possible electron microstates in the vicinity of a neuron.

And the best mathematical way to describe the sum of all probabilistic particle states, as we might recall, is either with a wavefunction or with a density matrix. Either formulation of quantum mechanics is applicable here, since either one can be used to describe all the possible energetic arrangements of a physical system. These mathematical tools will help us explain how the macro-scale computational unit – or the neural network as a whole – generates information as a physical quantity.

In this view, each cortical neuron is in a probabilistic state, dependent on the probabilistic position and momentum of all electrons in the vicinity, which define the voltage potential of the cell itself. Due to the random movement of electrons and the neural processes that rely on this random electrical noise, probabilistic particle states are likely to contribute at the level of cellular activity – and even network-level activity. That is because an electron can be either inside one cell or another, but not both. The state of every neuron in the network is resolved only when the state of every component electron is resolved. As a result, the state of the entire system must be calculated at once.

One could argue the only real uncertainty is due to the limits of precision measurements in the cellular-scale experimental systems we employ for studying neural networks – and once we improve our measuring capability, we will find that neurons act purely deterministically. Alternatively, we may realize that true eigenstates – and the forces affecting them – do affect neuronal signaling outcomes as we look more closely.

The undeniable existence of uncertainty at the quantum level suggests that we live in a non-deterministic universe. It is likely that we instead live in a world of probabilities – with each particle across three-dimensional space having some probability of existing in a given microstate (with a specific position *and* momentum) and some probability of the entire neural network containing *that set* of microstates (itself a macrostate).

When the probability of a microstate is *either* 0 or 1, the state is highly predictable and there is a low amount of information. But when a probability is somewhere *between* 0 and 1, there is a high amount of information and low predictability. Information is directly proportional to the uncertainty in the system – and the physical quantity of information held by a system is the sum of all possible microstates. Therefore, a system that is sensitive to the inherent uncertainty of particle states – and maintains that uncertainty for long periods of time without dissipating heat – can reliably sustain some quantity of quantum information.

So, we must choose how to describe this particular system. The macro-state of a cortical neural network, with each neuron firing an action potential or not, represents an instantaneous digital summary of the information content held by the brain. That is a fair way to describe the system. But taking into account all of that uncertainty at the quantum level may prove useful.

While the state of each neuron – firing an action potential or not at any given moment – is essentially a digital signal, representing an instantaneous readout of each cellular state, the sum total of all component microstates provides an analog signal, comprising the total information content held by the neural network. Of course, *both approaches represent the same underlying dataset*. Both provide a snapshot of the present condition, with each neuron in the entire neural network *at 0 or 1*, or each having some probability of reaching action potential threshold that is *between 0 and 1*.

The non-deterministic nature of electrons – which contribute to the cortical neuron voltage state – is the key here. The amount of information held by a physical system is given by its measure of uncertainty. Since cortical neurons retain sensitivity to the probabilistic movement of charged particles, the amount of information produced by a cortical neuron is truly dependent on the probabilistic state of each electron in the cell's vicinity.

Here, I am proposing that quantum uncertainty is actually harnessed by cortical neurons to create some physical quantity of information. In this view, neurons are not classical, binary, computing units – always in an off-state or an on-state – instead, they are qubits, which calculate the *probability* of switching to an on-state from an off-state.

In this view, the sum of all possible microstates for a cortical neural network (the total information held by that system) is far larger than that of a standard computer chip bearing the same number of binary computing units. Also in this view, the output of a cortical neural network is truly non-deterministic, with probabilistic events actually contributing to – or even fueling – the computational process.

CHAPTER 15
A mechanism for top-down processing: information compression affects neural activity

At the quantum level, reality can only be described in probabilistic terms – mathematical probability is not just a way of describing reality, it is reality. Here, I am proposing that biological organisms make use of probabilistic particle states in order to effectively perceive the world and direct behavior. In other words, consciousness can be explained in terms of probabilistic computation! In the previous chapter, I explained how cortical neural networks physically create information. In this chapter, I will explain how that information is compressed, with the results of that non-deterministic computation directly affecting neuronal signaling outcomes.

Essentially, incoming sensory data is physically integrated into a cohesive quantity of information in a cortical neural network, as upstream signals are integrated with random electrical noise. Individual cortical neurons sit at the threshold for firing a signal, integrating those upstream signals and random electrical noise to 'decide' whether they send a signal onward. The probabilistic state of every individual electron contributes to the voltage state of multiple nearby neurons, and so the entire neural network is suspended in a probabilistic state. As the system interacts with its surrounding environment – another particle system – consistencies emerge, allowing certain system states to become more likely. Interestingly, the states that best match reality will be most consistent, and so those system states will be favored. That is how a cortical neural network can actually encode the state of reality – by finding these correlations or consistencies.

The emergence of consistencies and redundancies in the dataset then naturally reduces the uncertainty across the distribution of possible states, with the most likely state becoming favored. Patterns or consistencies are recognized across the system, in a cascade of constraints. The *compression of information* that results from this process is correlated with the extraction of *meaning* from the dataset, as the neural network identifies the most likely true state of the environment.

The introduction of constraints, during an interaction, reduces uncertainty. This reduction in probability density – a reduction in entropy – must be converted to energy. This return of energy to the system after information compression is a thermodynamic requirement. In accordance with the Hellman-Feynman theorem, the reduction in the probability density – the collapse of the wavefunction – has a distinct physical effect on the atoms whose uncertainty has been reduced. Each atom experiences a boost in angular momentum, due to a redistribution of charge, as electrons are rearranged. In that situation, some atoms may be hurled across the nearest neural membrane, with that ion leak prompting neurons to reach voltage threshold and fire a signal.

This theoretical framework asserts that only a subset of neurons will be susceptible to the top-down signal – and a relevant subset of neurons at that. The state of the neuron at the moment of wavefunction collapse specifies its response. In other words, there are some states that are stable and other states which can be nudged. The wavefunction collapse is essentially a reduction in entropy throughout the system, with this system-wide event forcing a shift in energy distribution across every component ion in the system. That event could easily prompt additional ion flux across each neuronal membrane. Yet not all neurons would be affected by this 'nudge' – only neurons close to threshold would be prompted to fire an action potential (Figure 12).

Neurons already approaching action potential threshold (due to upstream signaling) exist in a temporal window sensitive to a nudge. If there is a system-wide computation, then a whole lot of cortical neurons sitting at action potential threshold should end up firing at the same time.

I propose that collapse of a network-wide probability density – a reduction of *entropy* – prompts simultaneous signaling outcomes in sparsely-distributed neurons across the network.

Because energy must be physically expended to create entropy, and energy is always conserved, the energy expended to create entropy simply cannot be lost. The first law of thermodynamics states that energy cannot be created nor destroyed, and so the energy expended to create entropy must exist in some form. As constraints emerge and the probability density is compressed, this entropy must be converted back to free energy and released into the system. The resulting release of energy, proportional to the decrease in entropy, drives ion movement in the direction

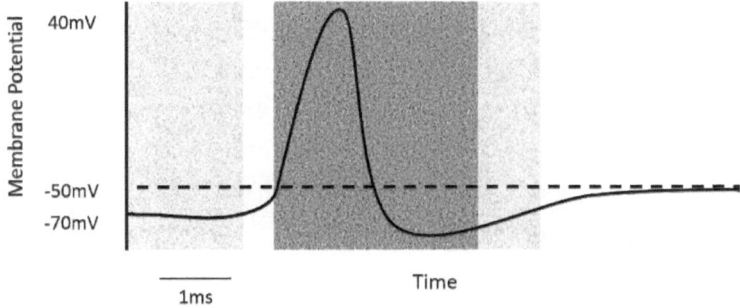

Figure 12. This theoretical framework posits that non-deterministic computation is possible, and naturally leads to the compression of information entropy. This physical information processing prompts instantaneous signaling outcomes in a range of sparsely-distributed cells across a cortical neural network. Neurons that are approaching action potential threshold should be most sensitive to this event. This figure demonstrates the stages of the action potential where free energy release has the greatest power to boost the neuron above threshold. The cell is *insensitive* to additional charge flux during the action potential and hyperpolarization period (shown in the region with a dark grey background). The cell is *minimally responsive* to additional charge flux at resting potential (shown with a light grey background). Yet when the cell is near voltage threshold (shown in the region with the white background), it could be pushed to fire an action potential with just a small additional charge flux. A cortical neuron approaching threshold could be very sensitive to a nudge delivered at the correct time, when the cell is in up-state, sensitive to random electrical noise in triggering a signaling outcome.

favored by the electrochemical potential, with the reduction of uncertainty driving signaling outcomes. Neurons that are near action potential threshold, sensitive to increased charge flux, are prompted to fire action potentials – simultaneously – upon wavefunction collapse, as information is compressed and the probability density for every component microstate is reduced.

This theory of system-wide computation provides a potential mechanism for the synchronous firing of sparsely-distributed neurons that is observed. In short, I argue that synchronous cortical neuron firing is a result of *information compression*, as information is parsed for *meaning* or *consistency with reality*. This computational process is therefore proposed to underlie sparse coding and the perceptual binding associated with that synchronous neural activity. Essentially, since information is generated and compressed at regular intervals, synchronous firing must occur at regular intervals, forming oscillations.

This oscillatory activity is observed in cortical neural networks. Some computational models suggest that neural oscillations can arise spontaneously through stochastic processes, with the system forming resonant frequencies in coupled subnetworks and then, with further recruitment, strongly-coherent oscillations across the entire network. However, this effect is primarily observed with epileptic-like spontaneous activity; normal high-frequency oscillations have a limited ability to recruit broader regions of the network. These findings suggest that classical mechanisms cannot produce the distinctive synchronous neural activity that is observed in the healthy brain [108].

Some researchers have hypothesized that neural networks may just form oscillations naturally – with order arising from stochastic activity – then somehow spontaneously cease further recruitment before the synchronous activity becomes epileptic. Alternatively, neural oscillations may be evidence of top-down computational processes. The collapse of alternative eigenstates is proposed here to occur during perception, with surprising (but nevertheless acceptable) data correlating with the reduction of alternative possibilities and the acceptance of a single reality.

This information processing event should therefore manifest as both a change in the mental state *and* a change in the total entropy throughout the brain, as a given percept is accepted and high levels of uncertainty are systematically reduced to defined particle states. As each computational cycle is completed, the most likely state for each electron is assigned, and some atoms may find themselves inside a neuron rather than outside. Such events may nudge neurons that are already near threshold to reach action potential threshold, leading to synchronous activity.

In support of this idea, studies suggest that a lower-probability yet acceptable sensory event is associated with greater neuronal synchrony in areas of cortex which integrate sensory modalities. For example, a shape with non-coherent visual features (such as a square plaid pattern), which nevertheless moves as a single object, will evoke greater neuronal synchrony than similar-size shapes that contain coherent visual features [109].

Therefore low-probability events which are nevertheless 'believable' (e.g. a visual stimulus that appears to contain a set of objects, yet moves as a single object) provoke high neuronal synchrony; the resulting network-wide synchronous firing event binds perceptual stimuli and promotes the formation of memory. The study above demonstrates that surprising-yet-consistent events are associated with greater neuronal synchrony; this may indeed reflect a quantitatively larger probability state collapse.

A related hypothesis, posited by Christoph Herrmann, suggests that high-frequency oscillations reflect the neural process of matching sensory input with memory [110]. If so, stimuli which are harmonious or which match the existing contents of the mind should evoke a greater burst of synchronous gamma-frequency neural activity, compared to non-harmonious or unfamiliar stimuli. The plaid-pattern study described above supports this hypothesis, as do studies which find stronger gamma synchrony in response to words versus pseudowords [111] and to faces versus abstract unidentifiable shapes [112]. It therefore appears that synchronous cortical neuron activity is indeed associated with the recognition that seemingly low-probability events are actually true, particularly in cases of perceptually consistent stimuli and familiar (remembered) stimuli.

Given these studies, it makes sense to think that integrating new information might reduce the probability density of the entire neural network, as low-probability events are discarded and high-probability events are accepted, and the information is encoded. In this view, a system-wide computation corresponds with a physical compression of information entropy – as multiple eigenstates collapse into a single computational outcome within the neural network, each electron takes on a defined state, and any neurons which have reached action potential threshold fire a signal, leading to a coordinated burst of neural activity across the entire network.

In short, a system containing fundamental states of uncertainty has incomplete causal determination, and that may provide an exploitable opportunity for an information-processing system to process data about its environment in a very real, physical way. To do so, energy must be expended to create a set of possible system states, but that energy is recovered during information compression. This theoretical framework asserts that biological systems obey the laws of mechanics and thermodynamics to achieve efficient, exascale information processing. Let's delve into these biophysics some more, to find out how consciousness naturally emerges from this computational process.

CHAPTER 16
The connection between energy expenditure and information processing

In order to explore how information is generated and processed to build a meaningful understanding of the world, it is necessary to consider how physical systems expend energy to do work. By investigating the brain as a physical, thermodynamic system, we can discover how our brains generate a streaming perceptual experience, develop a cognitive model of reality, and allow us to act on that understanding, all while encoding this information in the very structure of the cortical neural network.

The integration of information across the brain suggests an elegant solution to the binding problem, with information *bound in time* across the neural network – thereby providing a physical basis for a cohesive stream of multisensory perceptual awareness. Here, we will explore how electrochemical cells produce and process information over time – specifically, by considering how neurons dissipate energy into heat, into physical work, and into stored quantities of information.

The human brain is a thermodynamic system. Of course, every particle system is a thermodynamic system! But the human brain is a special kind of thermodynamic system, in two important ways. Firstly, it is far from equilibrium, actively gathering caloric energy from its environment. Secondly, it traps heat, then uses that thermal energy to do work. Thermodynamic systems like this have very interesting properties.

We regularly collect energy from our external environment by eating food. About one-quarter of the glucose we ingest goes to feeding our brains, and about one-fifth of the oxygen we breathe goes toward maintaining brain energy metabolism.

Despite being only 2% of our body weight, the brain uses up 20% of the body's energy intake – approximately 20 Watts, or 20 Joules per second. The brain does not release much energy either – only discharging 0.15 Watts of heat, or 0.15 Joules per second.

What is all this caloric energy being used for? This is an excellent question. Cells use caloric energy to do all kinds of work, from gene transcription to protein manufacture, to transporting key molecules across the cell to wherever they are needed. Cells also use energy to build lipid membranes, to repair cellular machinery, and to make raw materials for inter-cellular signaling. Neurons also have the unique characteristic of expending energy to build electrochemical potentials across their cellular membrane and remodel their synaptic structure to more efficiently transmit information to other cells in the network. All this work requires a lot of energy. And all this work must create entropy.

Thermodynamics allows us to quantify how energy is used. In fact, one simple equation calculates how energy is distributed across a biological system.

The amount of caloric energy entering the system must equal: the amount of energy stored in the electrochemical potential, plus the amount of energy expended toward work, plus the amount of energy expended on entropy, plus the total amount of heat (thermal energy) leaving the system. Because the first law of thermodynamics states that energy can be neither created nor destroyed, but only converted, these four thermodynamic quantities (stored energy, work, entropy, and heat) must perfectly sum up to match the total amount of energy entering the system.

Entropy is the quantity of energy entering the system that cannot be directed toward *heat* or *work*. It is the inefficiency of the system – all of the possible system states that will never be actualized.

So now we can calculate the energetic efficiency of our favorite thermodynamic system. The amount of caloric energy entering the brain is approximately 20 Watts or 20 Joules per second. The amount of energy leaving the system as heat is approximately 0.15 Watts or 0.15 Joules per second.

ATP is the molecular currency used to accomplish work in biological cells, which can be used to estimate the amount of work being done. Estimates of ATP turnover suggest that enormous amounts of energy are dedicated toward cellular work – transcribing DNA into RNA, manufacturing proteins, converting raw materials into signaling molecules, transporting these molecules to distant reaches of the cell, repairing machinery, remodeling synaptic connections, and pumping ions across the plasma membrane to set up the electrochemical potential. Indeed, most of the energy entering the system ends up being efficiently used to do work.

If we do the math, we find the quantity of entropy created by the human brain is tiny. It appears the brain is very nearly 100% energy efficient. There is almost no uncertainty or inefficiency.

Somehow, a biological system whose primary job is to parse information creates almost no physical information at all.

Such incredible energy efficiency, combined with the exascale computational capacity of the brain, is unimaginable for any engineered system currently within our technological capability. The brain must be doing something pretty interesting to appear like it is not producing any entropy.

The brain of course must produce huge amounts of entropy. For example, we can consider the amount of Shannon entropy produced by a biological neural network: The larger the number of neurons in the network n, the greater the quantity of entropy in the system, because each neuron could be either in an on-state p_{on} or an off-state p_{off}. However, any active neural network with n greater than 1 has *some* quantity of information. A biological neural network – with 86 billion cells, each in an on-state or an off-state – creates *a lot* of Shannon information.

What about Gibbs entropy? There is a lot of this too, since there are many possible microstates for this physical system. Every sodium ion in the brain has some possible relationship with every neuron in the brain, with some probability of being inside a given neuron, p_{in}, or outside that neuron, p_{out}. And the state of each ion, p_{in} or p_{out}, is determined by the state of each of its electrons, which are of course influenced by the local electrical field generated by the neuronal membrane potential. There are a lot of dynamic variables at play here, and as a result, an awful lot of possible system states! Why then is there so little net entropy being produced by the brain?

How can a biological neural network achieve such incredible computational power, while remaining so energy efficient? The entropy estimate given here is not a mistake or a fluke; studies have demonstrated that neurons in the cerebral cortex are strangely energy-efficient, compared with other types of neurons, which are only around 50% energy-efficient [113,114].

I propose that the human brain *is* producing a lot of entropy, but it is then able to compress this entropy *to extract meaning* – that is, *to extract a signal from the noise*. This process is possible because this highly thermoregulated system maintains tight temperature control. In this specialized system, thermal energy is not dissipated, but rather used to fuel *computational work*.

By parsing information, the system selects a single actual state from a large probability distribution. The realized system state will be the one most thermodynamically favored in the present context, as correlations between the present state of reality and the present state of the encoding structure are found. In other words, neurons in the cerebral cortex are *compressing entropy to do work*. The result of this natural thermodynamic process is a system-wide computation, which encodes the present state of the local environment. As entropy is compressed, free energy must be released, driving signaling outcomes.

Now, very astute readers will be asking the question: Doesn't compressing entropy – essentially reducing the disorder of a system – violate the second law of thermodynamics?

The second law of thermodynamics states that entropy always increases over time. Surely the reduction of entropy during a thermodynamic computing cycle disobeys the second law!

It turns out, the second law of thermodynamics is not a physical restriction like the first law of thermodynamics. It is more of a probability that systems will become more disordered over time, rather than more ordered, simply because there are many more disordered states than ordered states! So, it is a lot more likely that a random system will be found in a disordered state than in an ordered state.

The second law of thermodynamics is not an inviolable law of physics, *since entropy can be reduced locally*. This is the case *only* in non-equilibrium thermodynamic systems that trap heat to do work – such as biological systems, which tend to become more ordered over time [115-117].

Over the course of a single thermodynamic computing cycle, some energy will always be lost to entropy, so entropy *is* increasing over time, in accordance with the second law. But *within* a computational cycle, entropy is reduced whenever a large probability density is compressed into an actualized system state. That uncertainty cannot be completely reduced – there will always be some uncertainty regarding the syntactical relationships of any semantical truth statements realized during a computational cycle. This is a measure of the inefficiency of the system – the amount of uncertainty that cannot be reduced. Once uncertainty is reduced as much as possible and information is compressed, the system state is defined, and a new interaction kicks off. Uncertainty is instantly renewed as soon as the system state is defined, generating a fresh quantity of entropy.

What about the first law of thermodynamics? This law states that energy cannot be created or destroyed. Does the proposed process of information generation and compression violate the first law of thermodynamics?

The answer to this important question is no. The first law of thermodynamics is indeed an inviolable law of physics, and it is respected in this theory. In fact, the theory only works because of the first law of thermodynamics! The amount of energy entering the system must be perfectly balanced with the amount of energy stored, the amount of work done, the amount of entropy produced, and the amount of heat released, in order to zero the account in each and every computational cycle.

During quantum information processing, energy is never created or destroyed; it only changes form. The amount of *predictive value* or *consistency* or *meaning* in a quantity of information is actually defined and limited by the first law of thermodynamics. The quantity of predictive value extracted must *exactly* equal the quantity of information compression, the quantity of entropy reduction, the quantity of free energy released, and the quantity of work that is done to encode that predictive value into the neural network. There is another limit given by the first law of thermodynamics – the total amount of information held by a system, from which predictive value can be extracted, must be no greater than the total amount of incoming caloric energy minus the total amount of outgoing thermal energy.

Within these limits, however, a lot of computational work can be done. And a lot of *meaning* can be gained, as a neural network finds a compatible state with its local environment, encoding that external reality into its internal system state.

CHAPTER 17
Non-deterministic computation yields semantical and syntactical statements

What happens when our brains produce entropy? They expend energy to produce a set of possible system states. The quantity of entropy increases as we gather data from the senses. As that information accumulates, we can gauge the level of certainty we have about a given stimulus. That is, we gauge the amount of predictive value in that quantity of information. By extracting predictive value – or the *consistency* in a dataset – we compress information. This is a process of thermodynamic computation – we expend energy to create information or uncertainty, then we recover some of that energy when we compress information or reduce uncertainty [118]. As *meaningful patterns* are discovered within the dataset, the set of all possible system states is reduced into a single actualized state, and all other unrealized states fall away. The wavefunction collapses, the probability density reduces, information entropy is compressed, and some of the free energy that was expended to generate entropy is returned to the system.

The brain is a thermodynamic system that collects energy from its local environment, and uses this energy to encode the state of its local environment, by acting as a net heat sink. This system produces almost no net entropy. Instead, it achieves highly accurate exascale computation with incredible energy efficiency – with encoded information paired with qualitative information content. There must be some mechanistic explanation for how the brain operates, which accounts for the *representative content* we experience and the *meaning* we find in that information.

The voltage state of each neuron is related to the probability of ion flux across the neuronal membrane – and the position of each ion is defined by the probable state of its ensemble of electrons. *If the uncertainty in electron position, momentum, and energy state is sustained in the presence of a constantly changing electrical field,* these probabilistic states will contribute to the probable location of entire ions and the probable voltage state of each neuron.

In such a case, the best way to model the behavior of neurons, and ion flux across the neuronal membrane, is by representing the probable state of each component electron with a density matrix or a wavefunction. That allows us to quantify the total amount of information entropy in the system – and to describe how neurons calculate the probability of sending a signal.

The system cannot resolve its state on its own. It must interact with another system to do so. In the absence of a separate system to interact with, a single system on its own will just expand its number of possible states. If there is a separate system, and there are interactions occurring between the two systems over time, the two systems are forced to select a mutually compatible state from the combined probability distribution. This process defines the system state, and a new probability density emerges.

For a system that encodes data about the external environment, by gathering sensory data, the external perturbation which is informing our system state is *reality*. The other thermodynamic system with which we are interacting is our local environment, our surroundings – our reality!

Photons of light hitting our eyes and sound waves hitting our ears inform us about the state of our environment. These data are converted into electrical signals, and contribute to neuronal firing patterns across the brain.

In each moment, a whole range of system states is possible, with some level of uncertainty in both the actual state of reality and the pattern of neural activity which encodes that reality. Was there a bright light and a loud noise, coming from that direction? If the stimulus was particularly bright or particularly loud, we might be relatively certain that it occurred. If the stimulus is faint, we are less certain. The level of *certainty* we have that a particular stimulus occurred is encoded in our neural activity, along with the qualitative aspects of the stimulus.

Within a computational cycle, we may expand the number of possible system states, then reduce these states by identifying the most likely system state – the one that is most compatible with the interacting system. This allows us to do two things: Firstly, we reduce the quantity of entropy by selecting a single actualized state. That reduction releases free energy into the system, providing the energy needed to remodel the structure itself, to encode what we have learned. Secondly, we develop a semantic truth statement about the world. We have taken a set of possibilities and *extracted some meaning from the dataset by compressing the information*. Yes, there was a bright flash of light and a loud noise, coming from that direction! That happened.

We may be uncertain whether this event occurred, but there is a potential cost to ignoring it. We may decide to expend energy to verify these data, turning our heads toward the stimulus. In doing so, we can collect more information. What was that? Did a bomb just go off? Or is it the start of a fireworks display, which we have been looking forward to watching all evening?

Our expectations play a role in our level of certainty that a light or sound or other stimulus combination occurred in the way we perceive it. In fact, our expectations *for what stimuli we consider likely to occur in a given context* will actually affect what stimuli we perceive [119]. And these expectations are formed by prior experience. If we watch the fireworks every year – on the same day, in the same town, with the same people, we are unlikely to even consider that a loud noise and bright flash on that evening could be anything else but fireworks. If, however, we have not felt safe for a long time, we may have trouble imagining that something innocuous and even delightful might be happening.

Our expectations help us to gauge certainty about the stimuli in our environment [120]. It is easier to perceive something we have a high expectation of perceiving, and harder to perceive something we do not expect at all. When there is a great deal of uncertainty, we have to expend more energy to expand the probability distribution, in order to accommodate those unlikely events.

Once we have identified a semantical statement about reality – there is a bright flash and a loud noise – the next question is, what does this stimulus signify? Does this combination of events have any predictive value for events to follow? Is there *further meaning* to be gained? If the stimulus combination is already familiar, we may give it little attention. If it is unfamiliar, we may give it a lot of attention, *to find out what happens next*.

By holding semantic truth statements in memory storage, we may gain additional predictive value, at an energetic cost. That requires dissipating some energy toward information storage and toward collecting further information. But that is a worthwhile exercise! By integrating a sequence of semantical statements together, over time, we can hypothesize syntactical relationships between events. These hypothesized syntactical statements are essentially predictive models of the world. This is how we build an understanding of cause and effect – by observing the world and how reliably events tend to follow one another.

In other words, once we have figured out the current state of reality, there is still more potential predictive value to be gained. As a result, we do not compress entropy completely in a single computational cycle. We store that entropy, and we integrate it over time with other possible system states. When that happens, we predict a syntactical relationship between perceived events.

The realization of this additional predictive value also reduces the quantity of entropy held by the system, releasing free energy into the system to do work. This work can involve remodeling our neuronal synapses, to encode that newly-learned predictive value. In this way, we grow to understand our world. And that is what consciousness does – it gives us the ability to perceive our reality, understand how it works, and navigate within it.

For example, we may realize that bright flash of light and loud noise means a bomb is exploding in the area. We might want to take action, given this survival risk. In the future, when a similar combination of visual and auditory stimuli present themselves, we might predict a similar risk and take similar action. Memory, encoded into our synaptic connections, provides us a shortcut to interpreting newly-arriving information.

But our hypothesized semantic statements about reality may not actually be true – they may be distorted by false information, prior beliefs which interfere with a full and accurate perception of reality, and the inability to observe the whole of reality, due to the limitations of the sensory apparatus.

And yet, we are able to perceive reality! While our perception may not be complete and fully accurate, it is still enormously valuable. Yes, only a small range of the electromagnetic spectrum can be sensed by our eyes, and only a small range of mechanical sound waves can be sensed by our ears. There may be entire categories of events in the universe that we cannot know about, because we have not evolved senses to collect these data. But the fact that our knowledge of reality is incomplete does not mean there is no reality or that we cannot observe it. By contrast, we are well equipped to learn about our world, and doing so helps us to better navigate within that world. It would be wise for us not to deny reality, but to do our best to discover it.

The sequences of semantic truth statements, built into predictive models of cause-effect relationships between events in the world, may not be perfectly accurate either. A predictive model that is accurate in one context may simply be inaccurate in another context; the ability of the system to recognize the difference is determined by how effectively the neural network structure had been remodeled to encode that information. If the system remains adaptable, not completely ordered, more nuanced complexities can be noted and more predictive value can be gained over time, at some energetic expense. Holding untrue semantic statements as true – and attempting to build syntactical relationships between them – is equivalent to computational inefficiency. It wastes time and energy resources.

It is simply not useful to form internal mental models of the cause-effect structure of reality based on false percepts.

By applying these principles in practice, we can consciously choose to expend energetic resources to expand our perception, embracing data we have previously ignored. We may also choose to expand the number of mental models available to us, existing in a productive period of uncertainty that permits us to deliberate between several paths forward. And we may consciously select the behavior that is predicted to have the best outcome, in terms of achieving our goals. By tackling each stage of information processing, we can systematically achieve conscious perception, cognitive understanding, and volitional behavior.

In these past two chapters, we have discussed consciousness in terms of thermodynamics – with predictive value, consistency, or meaning being identified in some quantity of information. This computational process allows information to be *compressed*, as a signal is extracted from the noise.

Now, in the next two chapters, we will discuss consciousness in terms of the laws of holography and mechanics – with probability amplitudes physically distributed across complex planes – unobservable, but still very much part of our reality. These probability densities interact with each other, as we interact with our surrounding environment. This leads to a system-wide computation, with information being physically compressed as these wavefunctions or probabilities interfere. By studying this process of non-deterministic computation, we will be able to understand how exactly we construct a cohesive representation of qualitative information content, and how exactly we enact causation, acting non-deterministically in the world to achieve our goals.

CHAPTER 18
Cortical information processing is paired with representative information content

Neurons compute *information*, with the system state encoding useful data about the external environment. The action potential – an enormous sodium ion flux, causing a sudden change in voltage across the neuronal membrane, upon the detection of coincident upstream signals – encodes the qualitative nature of the external stimulus, the intensity of the stimulus, and some level of confidence the stimulus occurred.

Neurons also produce *entropy*, with a probabilistic system state that is dependent on the movement of charged particles. The generation of entropy is directly related to *the distribution of electrons near the neuronal membrane*. The probabilistic nature of cortical coding is key here, as neurons achieve a state change on the basis of both noisy charge flux and temporally coincident upstream events. Given this noisy coding, the amount of entropy produced by the brain is unfeasibly low. I propose that entropy is being compressed, as signals are extracted from the noise.

Once again, let's go through the proposed cyclical process of information generation and compression, step by step.

The first stage of the process involves *generating* information, by increasing the overall number of possible system states. As a neural network encodes the probabilistic movement of charged particles into the voltage state of each component neuron, the neurons themselves exist in a probabilistic state, with some likelihood of reaching action potential threshold and firing a signal. In doing this, a neural network creates both entropy (in a physical sense) and information (in a mathematical sense). These quantities are operationally equivalent, with both of them being proportional to the amount of disorder in the system.

The second stage of the process involves *reducing* that quantity of information - finding meaning in the dataset, or extracting a signal from the noise. Recognition of patterns in the data reduces or compresses the dataset, leading to a reduction in entropy or information across the entire thermodynamic system.

This two-step process is the foundation of conscious experience. While any thermodynamic system can physically generate information or entropy, only a system in which information compression contributes to the macro-scale system behavior will be conscious and capable of choosing actions based on this non-deterministic computation. Only a system engaging in this two-step process should be able to identify meaning or patterns in its environment, physically compress information, and use the newly-recovered energy to drive behavioral output.

But for the information to be *perceived* by the encoding system, there is another requirement beyond probabilistic coding: For information content to be *perceivable*, the encoding system must successfully encode information in such a way that it generates representational information content.

Interestingly, the connection between *encoding* information and *representing* information is well-described by wave mechanics, or the mathematics of holography. We can apply these principles to how information is encoded in the brain. But before we do so, let's go over the basic principles of holography.

In the case of a classical hologram, information is encoded on a two-dimensional surface and projected into a three-dimensional volume (Figure 13). The information is encoded by interfering waves. A reference beam interacts with an object beam, causing constructive and destructive interference to occur between the two waves. This interference pattern is encoded onto a flat holographic recording surface. As the reference beam hits that holographic recording surface, the encoded information is projected, restoring its original dimensionality.

The mathematics and physics underlying standard holography are well-described [121,122]. Interestingly, these laws can be applied to any system that encodes complex waves as *information* onto a polymer recording surface [123,124]. The dimensionality of the holographic reconstruction just increases in proportion with the dimensionality of the complex waves being encoded and the dimensionality of the encoding structure.

Now we can consider how information, taking the form of probabilistic particle states, can be encoded by a cortical neuron.

Each neuron can be modeled as a two-state quantum system, with some probability of being in an on-state or off-state after some time has passed. Furthermore, each sodium ion may be modeled as a two-state quantum system, with some probability of being inside or outside a given neuron after some time has passed. And the state of each ion is dependent on the dynamic electrochemical potential of every nearby neuronal membrane *and* the state of every component electron – each of which has a fundamentally uncertain position, momentum, and energy state. A cortical neural network is therefore best modeled as a physical system that retains sensitivity to probabilistic particle behavior in deciding the state of every macro-scale computational unit.

As cortical neurons near their action potential threshold, the probability of remaining in an off-state or switching to an on-state must be calculated. The state of every cortical neuron inherently depends on the probabilistic state of every nearby electron.

The state of every electron within this system is uncertain, best described as a set of probability amplitudes distributed across the x, y, and z 'spatial' axes – and another axis, ω, which describes the likely energy state or atomic orbital of that electron. That additional axis is perpendicular to the spatial dimensions comprising our observable universe. The probability amplitudes are essentially spread out along a complex plane, which is not directly observable to us.

These probabilities, or complex-valued wavefunctions, interfere with each other. The constructive and destructive interference is just like that we are familiar with, modeling photons travelling across the z dimension to hit a two-dimensional holographic recording surface (Figure 13A). These four-dimensional complex waves interfere in the same way, as they hit a three-dimensional holographic recording surface – the polymer membrane of a cortical neuron (Figure 13B).

The mathematics underlying these two processes are the same.

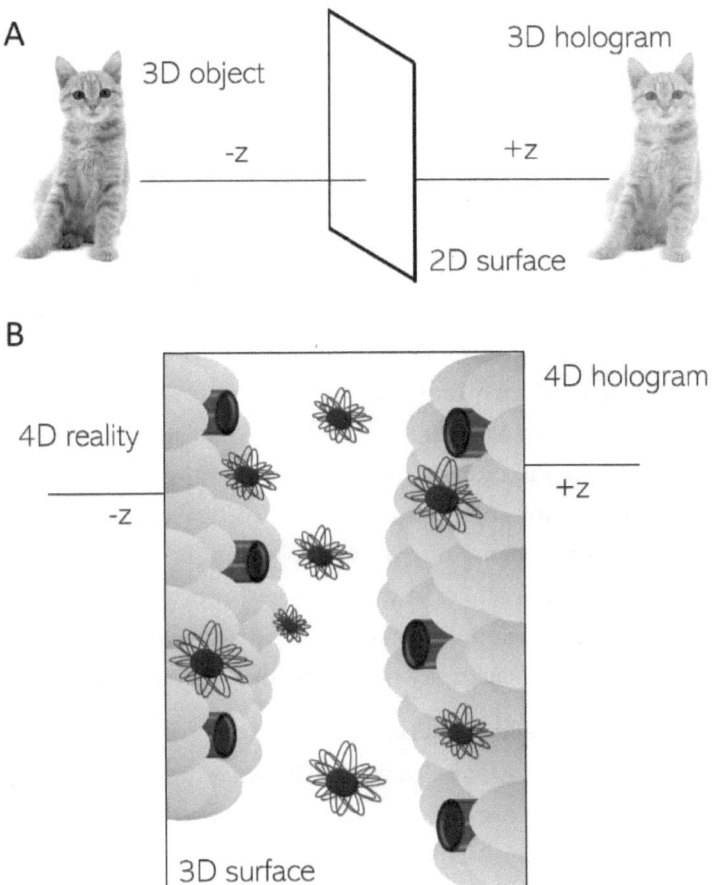

Figure 13. (A) A standard hologram encodes interfering waves on a two-dimensional surface and projects that information content into a three-dimensional volume. **(B)** Within the brain, the state of every component electron is uncertain, a set of probability amplitudes distributed across the x, y, and z 'spatial' axes and another axis, ω, representing the energy state or atomic orbital of that electron. These probabilities, or complex-valued wavefunctions, constructively and destructively interfere as they hit a three-dimensional holographic recording surface – the polymer membrane of a cortical neuron.

Just as information from three dimensions can be encoded on a two-dimensional surface and projected into a three-dimensional hologram [121,122], information taking the form of four-dimensional wavefunctions may be encoded on the three-dimensional surface of the neural membrane and projected into a four-dimensional hologram, representing the spatial location and intensity of stimuli.

Here, the immediately previous state for a particle will act as a reference beam, and all possible trajectories for that particle will act as object beams. Interference between these complex waves, on the neuronal membrane surface, allows information to be both encoded and projected. The projected information content represents data about the local environment, gathered by the sensory apparatus and encoded into the cortical neural network. And so, the proposed holographic projection should correspond to a cohesive stream of multi-sensory information content, continuously updated with incoming sensory data, yet limited by the range and sensitivity of the sensory modalities available.

Once the interference between complex wavefunctions occurs, a system state is transiently defined. The wavefunction collapses, information is compressed, and all electrons in the system take on a discrete spatial position and atomic orbital. The arrangement of electrons, relative to every neural membrane surface, is assigned a defined state at some location in time. That instantaneous event encodes a multi-sensory percept, or a semantical truth statement about the local environment (Figure 14A). At that very instant, uncertainty again begins to evolve and probabilities again start to populate the complex plane. The computational cycle repeats, and another percept forms, just milliseconds later. In this theory, perceptual experience is a continuously-updated holographic projection of information content encoded by cortical neurons.

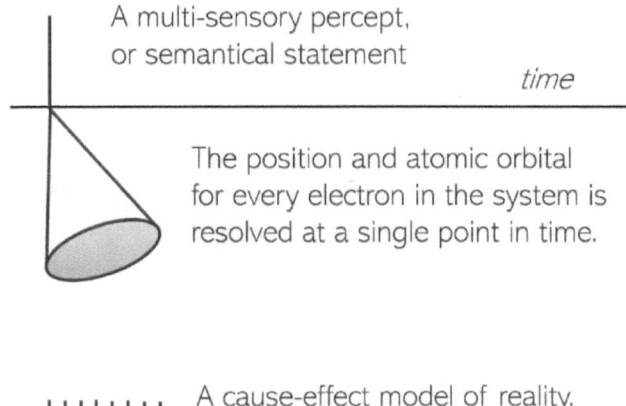

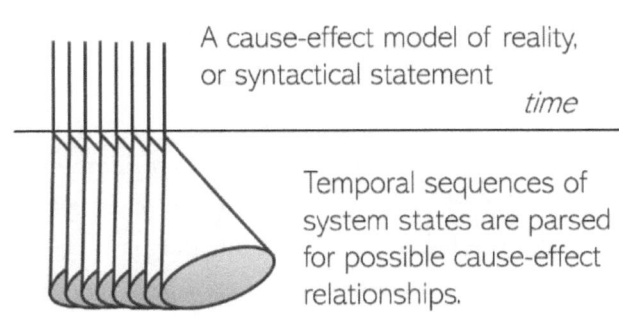

Figure 14. (A) As the wavefunction collapses or the density matrices commute, each electron is given some defined spatial position and atomic orbital, relative to every neuronal membrane surface. This defined system state only exists at a defined location in time. That instantaneous event encodes a multi-sensory percept, or a predicted semantical truth statement about the local environment. **(B)** These individual percepts are then integrated over the time axis, to relate a perceived event to immediately subsequent events. In this way, entropy still distributed along the time axis can be further reduced. Neural network states, each one paired with a multi-sensory percept, are integrated over longer timescales, to identify any hypothesized cause-effect relationship between events, or predictive syntactical truth statements about the causal structure of reality.

These individual percepts are then integrated over time. Since all uncertainty is not resolved in a single computational cycle – the relationship of a perceived event to immediately subsequent events remains uncertain – there are still some unresolved complex-valued probability amplitudes distributed along the time axis. This information entropy can be integrated over longer timescales, to create a hypothesized cause-effect relationship between events, or a syntactical truth statement about the causal structure of reality (Figure 14B). The physical integration of individual neural network states into temporal sequences should therefore produce a historical record of the information collected by the neural network over the course of a lifetime.

Since information is physically and mathematically defined as 'the sum of all possible system states', this quantity is both a mathematical description of the system state after some time has passed and an actual physical distribution of complex-valued probability amplitudes. But information that is *encoded* by a physical system is not necessarily *perceivable* by the system. The signals and noise must be effectively captured by the system and projected into a cohesive quantity of information content. This is achieved as the probabilistic behavior of electrons impinging on the polymer surface of a neural membrane naturally creates a holographic projection of information content. The hologram is encoded by the distribution of electromagnetic point sources, or complex-valued wavefunctions, which interact with each neural membrane, acting as a holographic recording surface.

This theoretical framework makes only two assumptions: 1) random electron movement contributes to signaling outcomes in cortical neurons, and 2) the cortical neural membrane is a charge-detecting polymer structure that meets the criteria of a holographic recording surface. Both of these assumptions are solidly supported by decades of neuroscience research.

Cortical neuron spiking patterns are highly probabilistic [103], while by contrast, the signaling outcomes of spinal neurons are highly deterministic [104]. It is well established in neuroscience that cortical neurons have unpredictable firing patterns [125,126], with signaling outcomes that are dependent on stochastic charge flux and spontaneous fluctuations in membrane potential [101,102,127]. In this new theoretical framework, it is that very process of harnessing the probabilistic movement of electrons that creates a system-wide probability density – physical information or von Neumann entropy – permitting the system as a whole to solve for the most likely state of every component electron in relation to every neuronal membrane surface. As a result, cortical neural networks are expected to yield representative information content and non-deterministic signaling outcomes, while spinal reflex circuits should not.

The membrane structure of a cortical neural network meets the structural and functional requirements to act as a holographic recording surface, thereby satisfying the second criteria. Classical holograms encode information about each local electromagnetic point source on a two-dimensional surface and project the rich, qualitative information content into a three-dimensional volume [121,128]. These physical principles can be extrapolated to any wave-like particle [123]. In this case, the putative holographic recording surface is a gelatinous phospholipid bilayer whose shape is described across three orthogonal axes: x, y, and z [129]. Physiologically, this surface is known to retain sensitivity to probabilistic particle behavior, encoding the trajectories of ions into the voltage state of the computational unit. Anatomically, the polymer membrane of the neuron meets the criteria for an ideal holographic recording surface [122,130]. These criteria include: 1) a charge-detecting surface comprised of synthetic or organic polymers; 2) a linear relationship between energy exposure and the amplitude of the reconstructed wave, thereby attaining signal fidelity; 3) a flat spatial frequency response, ensuring signal capture; 4) a large dynamic range, providing a good signal-to-noise ratio; 5) a high-quality and lossless material, affording efficiency in projecting the hologram; 6) sensitivity to very low energy exposure, achieving fine signal detection; and 7) protection from harsh environmental factors that could impact functionality. Interestingly, the cortical neural network structure also changes over time, as the gelatinous lipid structure ripples gently. As a result, this structure is able to encode information across the time dimension, as well as three spatial dimensions, since the encoding surface itself is defined across four axes. The membrane of a cortical neuron is therefore expected to encode, represent, and store information over time. Any information that is not compressed during a computational cycle will be stored along the time and energy axes. That information remains available for future parsing, unless it is dissipated as heat.

By making these two empirically-grounded assumptions, we find that perceptual experience naturally emerges from cortical neural networks. This approach readily differentiates between spinal reflex circuits (which are *not* correlated with conscious experience) and cortical processing of sensory inputs (where neural activity *is* associated with conscious experience). This theoretical framework shows how deterministically-acting, simple three-neuron peripheral circuits cannot produce mental representations, while cortical neurons can, because they retain sensitivity to random electrical noise. This theory asserts that subcortical processing in any sensory modality is insufficient for consciousness to manifest, because these neural circuits do not meet the two specific criteria necessary to generate a detectable holographic projection of information content.

It should be noted: the goal of this theory is *not* to disprove any previous neuroscientific research. The goal here is instead to understand perceptual experience, cognitive modeling, and non-deterministic signaling outcomes as naturally-occurring by-products of cortical neuron computation – thereby adding to the explanatory power of neuroscience.

This theoretical framework is called Conifold Theory because this whole process of thermodynamic computation is predicted to naturally form a high-dimensional shape called a conifold. Conifolds are thought to arise naturally in physical systems, as the arrangement of particles changes over time [131,132]. Conifolds obey the holographic principle – with a three-dimensional information-encoding structure in the observable universe; a four-dimensional information-streaming dimension, which represents the set of possible states of that particle system in the present moment; and an information-accumulating dimension, which integrates all of these individual system states over time to form the five-dimensional base of the conifold.

In this theoretical framework, a non-equilibrium particle system – which physically changes over time without dissipating heat into the external environment – creates a five-dimensional conifold structure that obeys the holographic principle, with information *represented* on a lower-dimensional boundary region and *paired* with a higher-dimensional volume of qualitative information content. The observable system encodes information, with a probabilistic arrangement of component particles across three spatial dimensions. The exact arrangement of component particles is uncertain, with regard to their spatial location and energy state. That integrated information – that is, all possible system states – is defined across four dimensions, as some set of complex-valued probability amplitudes which describe the likely spatial location and atomic orbital of every component electron. These probabilistic states of every component electron in the system are then resolved, yielding a transiently defined system state at a single point in time. These states are integrated into sequences, forming a historical record of all information that has been collected by the system over its lifetime.

And so, in this theoretical framework, a non-equilibrium thermodynamic system physically accumulates information by distributing mass and energy into possible configurations. This information is then compressed to distribute mass and energy into an optimal configuration, given the present context. Here, I argue that a cortical neural network not only encodes information, but also creates a cohesive stream of information content, and integrates that historical information content into temporal sequences, to act as a predictive model or reference for parsing new information. As the newly-arriving information interferes with stored patterns, the entire system compresses information through a system-wide computation. The release of free energy that is associated with information compression assigns all component particles in the system into some optimal energetic configuration. That system macro-state *correlates* with the present state of the local environment, and drives contextually-appropriate behavioral output within that perceived reality.

CHAPTER 19
Cortical information processing is paired with non-deterministic outcomes

This new theoretical framework has a lot of explanatory power: Neural networks expend energy to create information. Extracting any predictive value or *consistencies* in this information leads to compression. At the completion of a computational cycle – as the wavefunction collapses and the number of possible system states is reduced to a single optimal state – free energy is released back into the thermodynamic system. This burst of energy drives signaling outcomes and fuels productive work in the system.

The summed quantity of probabilistic particle states is the total amount of information or entropy held by the system. Importantly, this information is a physical quantity. According to the first law of thermodynamics, energy cannot be created or destroyed, so any energy that is expended toward creating information or entropy must continue to exist in the universe, even if it is unobservable. In a non-dissipative thermodynamic system, this energetic quantity can be recovered to drive computational work.

As the system interacts with its surrounding system, any correlations or redundancies between the two systems cause a natural compression of information entropy. The reduction of entropy must be balanced by a release of free energy back into the system [133-136]. This rule is called the Landauer principle.

While thermodynamics is the study of how energy is used to do work and create entropy, mechanics is the study of the physical processes underpinning this energetic distribution. To discuss mechanics, we have to discuss what happens to density matrices or complex wavefunctions – those probability distributions that are created whenever a system evolves over time.

In the previous chapter, we considered how the wavefunctions interfere constructively and destructively, on a polymer surface, to both encode and represent information. That process allows us to perceive reality as we are parsing information about it. Next, we can explore how this mechanical process actually affects causation – leading to signaling outcomes across the neural network and thus driving non-deterministic behavior.

Let's map out the probabilistic behavior of every electron in the system once again. The trajectory of every electron in the system is defined as a set of probability amplitudes distributed across five orthogonal axes – the x, y, and z 'spatial' axes, the time axis t, and another axis ω which describes the likely energy state or atomic orbital of that electron. These trajectories are described by probability amplitudes, distributed along complex planes.

After some period of time t has passed, the position of each electron is a scalar multiple of the Planck length, with respect to its previous position; the amount of time passed is some scalar multiple of the Planck time; and the energy state of each electron is some discrete orbital configuration. Since each electron within the system exists in a probabilistic state, with an uncertain trajectory, each electron is best described as a complex-valued probability density defined along five orthogonal axes, and the entire system state is a sum of all these probabilistic states.

What's important to note is that the number of possible states may contain some redundancies. Since any redundant particle states cannot co-exist, these probabilities melt away, reducing the total number of possible system states. That compatibility restriction helps to resolve the system state.

Each computational unit must calculate its most optimal state, after being perturbed by the environment over some period of time. But since the probabilistic state of each electron can affect the probabilistic state of multiple neurons, the system must be considered as a whole, with the entire system state only defined as every component electron state is resolved. All possible micro states must be integrated to create a complex-valued probability density, describing the set of all possible system states. The most probable neural network state emerges as consistencies and correlations are identified in the incoming sensory data, and thermodynamic constraints are realized across the network. As a result of these correlations and thermodynamic constraints, *the most optimal system state to encode reality will be selected.*

As correlations emerge between the system state and the state of the surrounding environment, the entropy of the combined system is reduced or compressed. Some probability amplitudes may become dominant, and these define the boundary region of the probability density. Now we must imagine the geometry formed by this computational process: Just as we *integrated* all possible system states to form a high-dimensional volume of probability, we can now identify the most likely system state by taking the *derivative* of the total entropy of the system. This computational process yields eigenvalues, or observables, for every particle on the three-dimensional boundary region of that complex-valued probability density.

As energy is redistributed across the system, and each electron takes on a distinctive position and orbital state, the system is transiently defined. Each neuron has reached its voltage threshold, or it has not; it fires an action potential, or it does not. That present moment, now defined, immediately becomes the past as a new probability state emerges for the system – with each neuron returning to its resting state, and each electron forming a new set of possible trajectories as a function of time. The entire process then repeats, in a cyclical manner.

In this way, a non-equilibrium thermodynamic system that traps energy and retains sensitivity to probabilistic particle states to gate a state change in individual computational units may undergo computational cycles of information generation and compression, simply by identifying correlations between itself and its surrounding environment. This computational process of 'detecting the state of reality' naturally leads to a redistribution of energy across the system. As a result, the outcome of this computation is an optimal system state in the present context, comprised of mutually compatible component particle states. The entire computational process is summarized in Figure 15.

Let's go over the computational process in a little more detail. This exercise will bring together some of the key concepts in neuroscience, thermodynamics and mechanics.

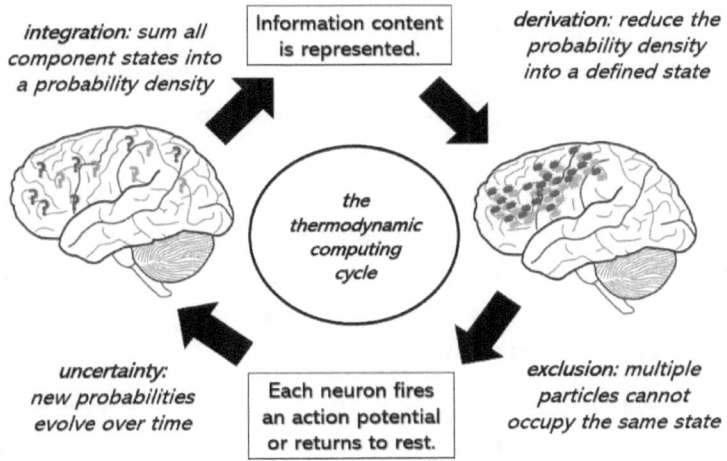

Figure 15. A schematic summarizes the thermodynamic computing cycle. Every electron in the system has some uncertain state, and the sum of all these component microstates generates a wavefunction, a density matrix, or a system-wide set of possible macrostates. The integration of all possible spatial positions and energy states across the system creates information, and taking the derivative of that probability density resolves the system into an observable state. The Pauli exclusion principle ensures that no two particles occupy the same state, thereby reducing the total possibilities with a Boolean satisfiability criterion. Ion states are defined; neuronal voltages are defined; and each computational unit fires a signal or returns to rest.

The probability of an instantaneous neural network state, with each neuron firing an action potential or not, can be calculated in relation to neural network states that previously occurred, as well as the intensity of the present perceptual stimulus. These factors, in combination, will guide the level of certainty that a stimulus has indeed occurred.

Each neuron has some probability p_x of firing an action potential. p_x is the present state of the neuron at resting potential, with some probability of firing an action potential. Neurons that are near voltage threshold have a rather high probability of firing, while those neurons which have just fired an action potential will be hyperpolarized and unable to send a signal again. This differential status was presented in Figure 12, with cells that are nearing threshold due to upstream neural activity being best placed to respond to a nudge, and cells which have just fired being relatively insensitive to any nudge.

This event should not only amplify coincidence detection from upstream signaling, but also create an instant internal snapshot of simultaneous activity across sparsely-distributed neurons, which can then be stored as a previously-actualized state, to be compared with subsequent system states. Any future replication of this state 'highly probable' compared with neural activity patterns which had not occurred previously. Previous system states, which represent familiar or consistent information, may contribute to probability collapse in the present moment, since these particular patterns of neural activity will be deemed 'highly likely'. In a callback to the previous chapter, this neural network state may be considered a 'reference' which clarifies the likelihood of the present state through constructive interference. A previously actualized state may have future predictive value!

As uncertainty is reduced, energy must be released back into the thermodynamic system, upon information compression [127-130]. Therefore, parsing predictive value and compressing entropy is predicted to boost the energy of electrons, with the amount of free energy entering the system proportional to the reduction in uncertainty. Every ion whose uncertain state is abruptly reduced will gain momentum, potentially being hurled into a nearby cell. In this way, a computational cycle may prompt ion flux across the neural membrane, allowing a subset of neurons to be nudged toward the voltage threshold for firing an action potential.

This theoretical framework proposes a specific hypothesis for how information processing might exert effects on the physical system which gave rise to the information set: the calculation of uncertainty regarding the existence of a perturbation – a light, a sound, or other stimulus – reduces a probability density into a single neural network state, affecting causation in the system.

If it is a common stimulus or stimulus combination, the neural network state encoding it is indeed probable, and there will be a large amount of certainty. Likewise, if it is an unusual but logically consistent state, it could also be deemed probable and acceptable. This reduction of uncertainty means realizing a set of eigenvalues. At this point, all unlikely states cease to exist, as a single reality is realized from the cloud of possible outcomes.

As this uncertainty drops away, there is immediately less information or entropy in the system. This reduction in entropy must be translated into free energy release, because of the first law of thermodynamics, which requires that energy is never created nor destroyed. This entropy is not dissipated; it is used to drive *work* within the heat-trapping system [133-136].

Any decrease in entropy must be converted into a proportional increase in the energy of those electrons whose uncertainty has just been reduced, during the calculation of likely system states. This event should lead to system-wide alterations in ion behavior – specifically a spontaneous acceleration of component particles that drives signaling outcomes and output behavior.

An increase in free energy, affecting electrons whose uncertainty has been reduced, may prompt some ions to be propelled across the neural membrane. This event may nudge some neurons to reach voltage threshold and fire an action potential. As a result, perceptual experience should correlate with synchronous activity across sparsely-distributed neuronal populations.

These information compression events should primarily affect the activity of neurons nearing the voltage threshold for firing an action potential. Such an event should only alter signaling outcomes in cortical neurons that are in up-state (e.g. neurons in the cerebral cortex which are actively maintaining a voltage state that hovers right near action potential threshold). The fact that cortical neurons expend a great deal of energy to coordinate and maintain this delicate state suggests that it is useful in achieving computation [125,137]. This theory asserts that coordinated cortical up-states allow probabilistic states to affect signaling outcomes.

The realization of eigenvalues for electron position, momentum, and energy state – in both free-floating ions and the atoms comprising the lipid membrane of the neuron – causes a shift in charge distribution and a boost in angular momentum, driving sodium ions into nearby cells. That in turn triggers a signal *only* in neurons approaching action potential threshold. As a result, information compression implements the computation into the neural network itself, coordinating the electrical activity of neurons and providing a mechanistic explanation for so-called supervening mental states in driving neural activity.

As some of the energy which had been converted to entropy is converted back to free energy, that quantity can be used to do additional work within the system – for example by remodeling synapses to reflect the information now held in memory, thus making these patterns of neural activity more thermodynamically favored in the future. In short, the free energy released during information compression can also be used to direct work toward remodeling the system into a more ordered state.

And so, by parsing predictive value to compress entropy in an energy-efficient computational process, cortical neural networks maximize computational power, minimize energy consumption, drive non-deterministic signaling output, encode information into memory through spontaneous self-remodeling, and accelerate from rest to achieve optimal yet non-programmed behavior. This theoretical framework therefore offers a mechanistic basis for thermodynamic irreversibility in biological systems.

Let's recap. The position, momentum, and energy state of every particle in a heat-trapping thermodynamic system is uncertain, and dependent on the actions of other particles. This sustained quantum state can be described mathematically, as a complex-valued wavefunction or a density matrix comprising all possible system states. This *physical quantity of information* is generated by probabilistic particle behavior, encoded into the neuronal voltage state, and paired with qualitative information content that is holographically projected into higher-dimensional space.

Each electron in the whole system must take distinctive stances in relation to position, momentum, and energy state. After all, no two particles can be identical – this impossibility is prevented by the Pauli exclusion principle. And as the state of each component particle is defined, the voltage state of every neuron is defined. Neurons fire an action potential or return to resting potential. The whole network encodes the state of reality.

But critically, the resolution of these probabilistic particle states are associated with a small shift in the charge distribution across each atom, in accordance with the Hellman-Feynman theorem. Due to these subtle shifts in charge distribution, accompanying the resolution of the quantum state, each atom accelerates – it changes its velocity and momentum. The extraction of a signal from the noise actually affects causation within the system.

Such a system-defining event, and its effects on atomic charge distribution, will be immediately followed by the formation of a new probabilistic state which evolves with time, in accordance with the von Neumann projection postulate. As time moves forward, the set of possible outcomes for the neural network will grow once again. Information will be cyclically generated and compressed, as long as the system operates.

Rather than letting quantum mechanical effects define the limits of our knowledge and actions, we may find it useful to accept the fact the universe is probabilistic, particles are wavefunctions, and neurons encode information in a physical way – in the form of high-dimensional probability densities. By taking this route, certain otherwise-inexplicable phenomena can be explained – from streaming perceptual experience, to the gradual formation of predictive models of the world over time, the spontaneous structural remodeling of synapses to store information content, and the volitional control of behavior.

Understanding and exploiting the laws of quantum information processing may not only help us to understand the brain but also may help us to engineer systems with non-deterministic signaling outcomes, local memory storage, general intelligence, and highly energy-efficient exascale computational power.

CHAPTER 20
Consciousness means exploring the world and exploiting the information gained

This theory asserts that any cellular unit which detects the coincident timing of ions crossing a physical barrier to gate a signaling outcome processes information in a real physical way, through the bidirectional exchange of free energy and entropy. Such cycles are predicted to occur whenever the uncertainty of an electron state contributes to ion behavior in the presence of a dynamic electrical field, as in a cortical neural network.

The resulting cycle of information generation and compression in our brains is both a thermodynamic computing process and a quantum computing process. It may therefore be reasonable to consider cortical neurons as qubits, which calculate the *probability* of switching to an on-state, rather than classical computing units, which are *only* in on-state or in off- state.

This theory asserts that consciousness is a by-product of quantum information processing, and it explains how biological neural networks achieve quantum computing in practice.

The extraordinary energetic efficiency of the central nervous system has certainly been noted among researchers who have questioned whether this competence is intrinsically linked to the production of information entropy [138,139] or the exascale computing capacity of the brain [140,141]. In this theory, energy must be expended to produce entropy – but rather than being irreversibly lost, the quantity of entropy is reduced as the most optimal system state is identified, and free energy is recovered by the system during that compression phase. Since unlikely or unfamiliar system states may hold predictive value, some quantity of entropy must be generated to accommodate these possibilities. Although this effort has an energetic cost, it may be worthwhile, by reducing uncertainty in the longer term.

Any information held in memory will be parsed for predictive value, as incoming information continues to accumulate. The more predictive value contained in that information, the more free energy is released back into the system. As the system works to encode the meaning it has gained, it remodels itself into a more organized state. As a result, disorder is reduced, less entropy is produced, and more free energy is available to the system. So, thermodynamic computing systems both *reduce* information entropy to maximize efficiency (by favoring previous patterns of neural activity) and *expand* information entropy to maximize potential predictive value (by engaging in exploratory behavior).

These two processes perform in opposition; the generation of information entropy is thermodynamically unfavored, but it is favored if the information provides predictive value. In combination, this rival cost function leads a neural network to self-organize into a highly ordered state and effectively compute the predictive value of any incoming data.

There is a useful analogy in the computing world to explain the dynamic between exploring the environment and exploiting previously gained knowledge. During unsupervised learning, algorithms will explore a space of possibility, which can be imagined as a sort of fabric. Finding a solution to a problem is finding a 'local minimum' or dip in this fabric (Figure 16). The algorithm takes steps across the fabric of possible solutions to a problem, and as it does so, it determines whether its own position is entering a dip.

Computationally, the gradient of the dip means that less energy is needed to reach a solution to the problem. And so, in every moment, the algorithm seeks 'gradient descent' or lower energy expenditure than was used for the last attempt. The thing about classical algorithms, running on classical computer architecture, is that once they find a solution, they do not willingly expend energy to exit that dip and find another solution. This is why humans are different from computers. We absolutely *will* explore our environment – trying new things, even challenging things, after we have found perfectly good ways to get food and other resources. However, we are of course also prone to exploiting our previously-gained knowledge – by sticking to the same routines and engaging in the same patterns of behavior we have already established, even when we know a better solution is out there. The well-trodden path is difficult to change, because it takes a lot of energy to exit a local minimum.

Getting enough energy through a healthy, nutritious diet is *necessary* to maintain effective information processing, but this contributing factor is not *fully sufficient*. In order to compute data efficiently, a biological or engineered system must have functioning hardware. After all, it is impossible to ride a bicycle

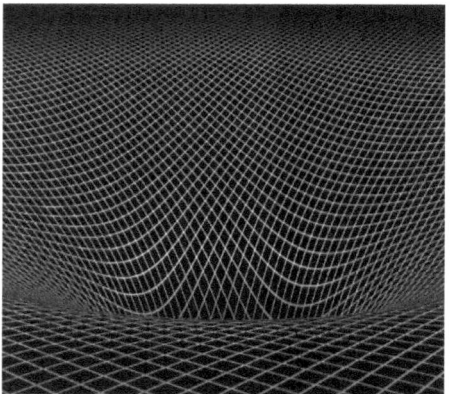

Figure 16. Shown here is a possible solution space. Neural networks seek out local minima, or dips in the solution space, to solve a given problem. Conifold Theory asserts that energy must be physically expended by a neural network in order to exit a local minima & discover new solutions.

when the gears are stuck. With the brain, functional hardware means having working ion channels in the neural membrane and efficient enzymes which produce all the molecular materials required by neurons to do their jobs. If these protein structures are *altered*, due to mutations in the genome, or if they are not *present in sufficient amounts* within neurons, due to changes in gene expression, then neurons may not be able to send messages to each other effectively. A loss of coordinated activity, due to biological impairments in signal processing components, may in some cases be corrected with pharmacological intervention.

On occasions when the structure and operation of the brain makes it extremely difficult to forge new paths – because the energetic expenditure is so great – people may benefit from pharmacological intervention to help them rewire their brains. Modifying the physical neural circuitry can help people to change their thought patterns and behavior. In combination with *a willingness to change* and *a great deal of effort*, these interventions can make it a bit easier for people to overcome addiction, depression, anxiety, and disorders of perception.

In general, understanding what consciousness is - and how biological neural networks work – can help us to better navigate our world and keep ourselves healthy.

Conifold Theory offers a framework for explaining consciousness in terms of physical information processing. Within this theory, the brain is an information-encoding structure, while perception and beliefs and memories are the information content encoded by that structure. Because energy must be expended to create information, we must physically devote energy to accurately perceive reality and to construct an understanding of reality.

In this view, our memories and beliefs are the information we have accumulated over a lifetime, and because they are useful to us, we do not easily dismantle them. So, whenever redundant information is identified, no additional energy has to be expended to accommodate it. As a result, these data are easily accepted. But whenever conflicting information is confronted, additional energy must be expended to accommodate it. As a result, it is often easier to discard these data as anomalous or incorrect.

If ignoring information or sticking to old pathways does present challenges for the individual to effectively navigate their reality, then it becomes necessary to spend additional time and energy engaging in conscious perception – forming new mental models of the world, articulating new goals, and choosing behaviors that have more positive outcomes.

In short, we will always take the easiest path to understanding our reality. If our current mental models work well enough, we will not expend energy to improve them. If our mental models are not predicting reality effectively, we will expend that energy to improve them.

Simply put, thermodynamic computing works on the basis of a rival energetic cost function. This rival energetic cost function involves maximizing the quantity of free energy available to the system in two different ways: *maximizing* entropy production, by continually exploring the environment to *gain* predictive value, and *minimizing* entropy production, by maintaining a highly ordered system state with *stored* predictive value.

This rival energetic cost function is achieved by exploring the world, then remodeling the neural network structure to encode the predictive model of reality that was acquired through this exercise. Both processes require expending energy to do work!

Exploration involves expending energy to discover the world. But it also yields predictive value, allowing information to be compressed and free energy to be recovered by the system.

The predictive value that was extracted during the computation is then encoded into the neural network structure, through an energetically-expensive process of remodeling the connections between neurons. But this synaptic remodeling process creates a more ordered state, thereby reducing the amount of entropy in the system and thermodynamically favoring the reoccurrence of sequences of neural network states that encode the meaningful information gained through experience.

So, organisms always aim to maximize free energy or reduce overall energetic expenditure. As a result of the rival energetic cost function, a maximization of free energy can be achieved by either the creation of entropy (through exploration, followed by the extraction of predictive value) or the reduction of entropy (through structural remodeling, to store that predictive value). As a general rule, biological organisms undergo repeated cycles of generating and reducing disorder – first producing and then compressing information – to achieve a more ordered and contextually-appropriate system state over time. Consciousness is about actively engaging with this process, by choosing how to expend energy – to acquire more knowledge, or to exploit the knowledge already gained.

And so, this theory predicts that consciousness is about computing information effectively. That means putting our neural circuitry to work, actively choosing how best to expend energy in order to meet our goals. There is no shortcut – we have to do the work. In practice, that means deciding between *exploring our environment* and *continuing down the well-trodden roads we have already constructed*.

We can always choose whether we expend energy, gathering new information and trying out new ways of doing things, or whether we save our energy by sticking to known paths and solutions that have worked well enough in the past.

If we want to change – if we want to create a healthier, more sustainable lifestyle, or a more peaceful and just world – we just have to put the work into finding new solutions.

CHAPTER 21
Consciousness as natural computation

The theoretical framework presented here does not postulate an exact equivalence between neural activity and consciousness, but neither does it posit the existence of a spiritual plane that is unamenable to physical law. Instead, this theoretical framework describes a natural process by which cortical neurons cyclically generate and compress *information* - with a cohesive stream of representational information content, non-deterministic but non-random signaling outcomes, and an increasingly ordered system state emerging from that computational process.

This theoretical framework is built from three independent mathematical models, which each demonstrate the same process of information generation and compression [142-144].

The first model links probabilistic coding in cortical neurons with a holographic projection of information content, by using the laws of wave mechanics. In this approach, complex-valued probability amplitudes or *wavefunctions* interfere constructively and destructively as they are encoded on the polymer membrane of each neuron. This model explains how incoming sensory data, encoded in a cortical neural network, yields a cohesive stream of representative information content that is exclusively accessed by the encoding structure and continually updated with newly-arriving data from the senses. This mathematical model shows how information could be physically encoded and projected in accordance with the holographic principle [144].

The second model formally employs quantum mechanics, to describe how eigenvalues are identified by taking the derivative of all possible energy states at all possible spatial locations. Here, observables, or eigenvalues, are the most dominant probabilities – the probabilities that are realized on the boundary region of a high-dimensional, complex-valued probability density. This approach explains how probabilistic coding within a cortical neural network leads to non-deterministic signaling outcomes, with quantum oscillations directly contributing to the voltage of the neural membrane. This mathematical model asserts that information is physically generated by probabilistic particle behavior, and these probabilistic events contribute to signaling outcomes in especially noisy neural networks like the brain [143].

The third model tracks how energy is expended toward entropy, and then recovered during compression, in terms of matrix mechanics and in accordance with the laws of thermodynamics. Here, a far-from-equilibrium particle system, called System A, interacts with its surrounding environment, System B. System A encodes the state of System B by extracting *consistencies* in the combined information. The identification of those consistencies causes information compression, releasing free energy back into

System A and triggering signaling outcomes that encode *meaning*. This third approach asserts that any information encoded in a cortical neural network is parsed for a best match with reality, yielding predictive models of both *the present state of reality* and *the overall cause-effect structure of reality*. This probabilistic form of computation should happen naturally in any net-heat-trapping thermodynamic system that is capable of noisy coding [142].

It is worth noting that all three of these mathematical models describe the *exact same* energy-efficient computational process of information generation and compression.

Is quantum computation actually possible? von Neumann and Feynman thought so. But it has proven difficult to achieve in practice – or even imagine what it means, in practical terms [145]. Yet we do know the thermodynamic processes of information generation and information compression are possible, and even well-described [118,146-148]. The fact that such computation requires physical probability densities and results in non-deterministic computation indicates that quantum computation is not only possible, but readily occurs at ambient temperatures.

What about these additional, complex-valued dimensions? Are they an actual part of our reality, despite not being part of our observable universe? Recent studies have shown these complex planes are indeed part of reality [149-151]. Even though we cannot observe them directly, these additional dimensions *must exist* for experiments to work out the way they do.

Simply put, it appears as if there is a reality, but it is a bit more complicated than we had previously imagined. In particular, there are violations of locality that are only possible if events are connected outside of our observable reality, on these complex planes. By postulating that particles interact probabilistically with each other, in higher-dimensional space, in accordance with the laws of quantum mechanics, we can make sense of how those probabilistic interactions become observable realities.

To make sense of how probabilities interact on a complex plane, outside of our observable reality, we need to think about reality in higher dimensions: We observe an object having only a single location across three-dimensional space at a single point in time. But that does not preclude an object from having another location at another point in time! If we think about all of three-dimensional space existing along another axis, the time axis, then we can start to imagine four-dimensional space-time – and start to appreciate how much of our universe is not observable to us.

Now, if we also think about every electron in every atom having some possible set of energy states, or atomic orbitals, we can think of those probabilities laid out along another axis. At every point in time, every location in three-dimensional space has some possible set of energies, and those probability amplitudes are laid out on a complex plane, perpendicular to our observable dimensions. Now it becomes clear that our universe may contain a lot more dimensionality than we had expected. What's really happening in the present moment is not defined, but rather a set of probability amplitudes described across five orthogonal axes.

Any system that acts probabilistically – with non-deterministic outcomes – is performing a physical form of computation. Any natural system, comprised of matter and energy, that grows into a more ordered state over time, with non-deterministic behavior, is undergoing a natural process of thermodynamic computation, by cyclically generating and compressing information. But only very specialized systems – comprised of individual computational units, each enclosed by a lipid barrier with the key properties of a holographic recording surface – will be able to effectively represent and perceive the information content being encoded.

These concepts are complicated and unfamiliar. The method of computation being proposed here is completely different from both digital computation (with binary computing units in either an on-state or an off-state at any given time) and current quantum computation (with spinor states being preserved at cryogenic temperatures to prevent any noise from disturbing the system). This method of computation takes into account a lot more about the probabilistic state of a particle (with spatial position, energy state, and time all being uncertain during the calculation) – and as a result, a lot more computational power is gained.

This method of computation specifically allows random noise to affect the computation – and this thermal noise, gathered from the external environment, *actually drives the computational process.* Here, probabilistic particle states interact with each other, on a complex plane, until the most optimal and compatible system state becomes thermodynamically favored. When that happens, things click into place, and then new possibilities start to evolve from there. This is a natural process by which a system checks itself against reality, regularly finds a way to stay compatible with reality, and exists in harmony with reality.

Building on this idea of harmony, we can consider an analogy: There can be any number of songs in imaginary time, but only those generating harmonics will become 'real'. These songs – the songs of our minds – are playing out simultaneously, in dimensions outside our observable reality. And the world that surrounds us is filled with songs in imaginary time as well.

It is when the songs in our minds interact with the songs of the world, on a complex plane, that harmonics start to form between the two. The songs of the mind that best match the songs of reality are the ones that will grow to dominate the din, with those harmonics forming a melody that is true and clear. That harmony happens whenever we grow to understand our world.

This process cannot be faked; the match has to be a good one for harmonics to emerge. If we cannot make sense of the world – if we cannot make out a melody – that means we need to think up more songs in imaginary time. By doing that work – by expending energy to create all of these possible songs – we can find one that matches the song being sung by our world. But we need to stretch our imaginations to meet reality – reality will not necessarily stretch to meet us.

Consciousness is the very process of engaging with reality, discovering what it comprises, and then encoding that reality into the brain. And doing that work is incredibly rewarding! Once you perceive the song, you can sing along, in harmony with the world and other people within it. And you may feel so good singing that song that you decide you want to jam, change the key, introduce a vibrato or some cool jazz notes. How well that effort works out will depend on how well you understood the song you were singing to begin with. If your new notes are too jarring, in the context of the ongoing music, then harmonics cannot form, and your voice will just get lost in the din. It's best to always keep in time with the music, and train your voice to always be in tune. With practice, you will be able to sing your own song – and when that time comes, you will know what song to sing.

Section IV
Exploring the Theory

CHAPTER 22
A point-by-point summary of the theory

Conifold Theory aims to explain what consciousness is and what it accomplishes. It takes a specific philosophical position, asserting that thought is indeed something categorically different from the matter and energy found in our observable universe – but it is still a physical process, operating within the laws of the physical universe. It builds on over 130 years of research in the field of neuroscience, contributing new insight while not disproving any of the important findings already made. The computational process underlying human consciousness can be explained in terms of thermodynamics, mechanics, or holography; each level of description is valid and in agreement. This theory provides valuable insight into the psychology of human behavior, and it also offers a generous gift to the fields of physics and engineering, by explaining how ambient-temperature quantum computation works in practice. Understanding what consciousness is, and what it accomplishes, may help us to perceive our world more accurately and navigate within it more effectively. Here, the key points of Conifold Theory are summarized.

1. Conifold Theory explains how consciousness works, in terms of a computational process that is tied to neural activity. The theory asserts that consciousness is something categorically different from neural activity; yet it also asserts this emergent property can be fully explained by physical processes.

As such, Conifold Theory takes a specific philosophical position with regard to the nature of the mind. In this view, our selves are not god-like avatars, beamed into our physical bodies. Nor is a neural network strictly equivalent to consciousness here. In this view, consciousness is a natural emergent property of information processing in non-deterministic neural networks.

This theory describes how a computational process gives rise to qualitative content, providing an improved explanatory framework to address the current impasse in consciousness studies. The theory addresses the seemingly immaterial basis of perceptual experience, the emergence of the self-concept, and the initiation of volitional action, in accordance with physical laws. As such, this new theory permits us to broach the age-old question of how mental states fit into the physical world.

In short, Conifold Theory asserts that perceptual experience is real, and contains the information content encoded by the brain; the concept of the self is real, and is an accumulation of information collected over a lifetime, in the form of beliefs and memories which form predictive models of the world; and finally, free will is real, with our actions only limited by the constraints of our own bodies and the physical laws of our universe. In this way, Conifold Theory takes a clear philosophical position on the nature of consciousness. It asserts that consciousness is a process of biological computation that permits us to perceive the world, understand the world, and choose our actions in the world.

2. Conifold Theory is well grounded in the laws of neurobiology. Yet this theory moves beyond classical assumptions to discover what sort of properties emerge from neural networks that engage in non-deterministic information processing.

We have made enormous progress over the past century in understanding the nervous system. Yet we have long lacked a theory for what consciousness is or how it operates. This deficit implies that we have been missing some major variable in our existing paradigm. At this point it may be useful to acknowledge the limits of classical physics, and specifically consider whether the features of consciousness may be a result of probabilistic neural coding rather than purely deterministic processes.

And so, Conifold Theory is completely consistent with empirical findings in modern neuroscience. It simply models, at a more detailed level, how interactions between electrons at the cellular membrane contribute to the voltage state of a cortical neuron. What arises from this new approach is a cyclical generation and compression of information entropy, as signals are extracted from the noise. This theory provides a plausible mechanism for the emergence of perceptual awareness as a holographic projection of information content encoded by the neural network. It shows how cause-effect models of the world emerge from sequences of percepts. And finally, it shows that compressing information provides the energy necessary to drive signaling outcomes and implement volitional action through the initiation of behavior.

Simply put, conscious experience is not possible without neural activity. That said, consciousness is something categorically different from neural activity. The *physical nature of this correspondence* is described in Conifold Theory, with the brain as an encoding structure and the mind as information content.

This theory asserts that consciousness is a natural phenomenon which allows us to experience a cohesive perception of the world, so that we may act effectively within the world. It is a natural phenomenon that is reliant on neural activity.

Indeed, thousands of studies have demonstrated how neural activity is required for perceptual experience. The experience of seeing colors and shapes is tied to neural activity in visual cortex [152,153]. The experience of hearing is tied to neural activity in the auditory cortex [154,155]. Not only are the characteristic patterns of neural activity paired with the *timing* of perceptual experience, but different patterns of neural activity across the cerebral cortex are correlated with different experiences [156,157].

Memories are also closely tied to neural activity. As we learn associations between a context and an event, or an important event that tends to follow a more innocuous event, the patterns of neural activity which encode that information become strengthened. The synaptic connections that encode reliable information will be reinforced, while synaptic connections that are not as useful may degrade. The nervous system is constantly remodeling itself as we explore the world and encode what we have learned [92]. These structural modifications favor the same patterns of neural activity to repeat – for example, a similar combination of electrical signals will fire throughout the brain when a person enters a familiar context. In this way, the brain physically encodes memories in its very structure.

Decision-making is also tied to neural activity. In higher regions of cortex, multiple sensory modalities are combined to influence motor output [158]. When neurons reach the threshold for sending a signal, this event corresponds to the *experience* of making a decision [119,159]. Our behavior is a result of that neural coding.

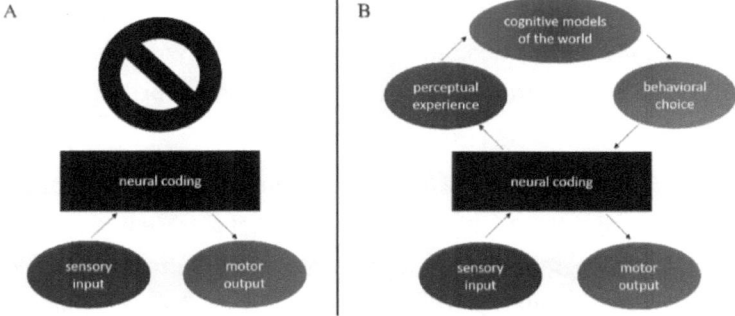

Figure 17. In classical neuroscience (A), sensory input is achieved by collecting data through the sensory apparatus: eyes, ears, tongue, nasal tissue, and skin. These data are processed within the central nervous system, with decision-making being a purely deterministic calculation resulting in motor output. Here, there is no role for perceptual or cognitive *experience* in decision-making or behavior. In Conifold Theory (B), patterns of electrical activity encoded within the neural network structure generate a holographic projection of information content, constantly updated in real time and tied to the neural network that created it. These neural network states are then integrated over time to form mental models of reality. The entropy reduction achieved by information compression releases energy, prompting non-deterministic signaling outcomes. Conifold Theory therefore offers a scientific explanation of conscious experience.

In summary, perception, memories, and decisions are encoded in our neural activity. And to a great extent, neuroscience has uncovered the correspondences between cognitive phenomena and the underlying neural network activity. However, neural activity is *not the same* as perception, memory, cognition, and decision-making. There is currently an explanatory gap, and this is where Conifold Theory proves useful (Figure 17). This scientific framework aims to explain the qualitative aspects of consciousness, and how these *relate* to neural activity.

3. Conifold Theory can be explained in terms of *holography*, with information encoded on the neural network structure projected into higher-dimensional space in accordance with the holographic principle.

Here, probabilistic particle movements are detected by the neural membrane structure. This three-dimensional surface is proposed to act as a holographic recording surface. Information is encoded in the pattern of interfering complex waves, as electrons interact in a probabilistic manner with every local neural membrane. The constructive and destructive interference between these complex waves, or probability amplitudes, yields a percept – a usable and meaningful signal, extracted from the noise. So, the encoding process generates a cohesive stream of information content, providing a qualitative perceptual experience of any incoming sensory data. In thermodynamic systems that persist over multiple computational cycles, without dissipating much heat, information can be stored. As information is stored by the neural network over a lifetime, that *accumulation* of stored information forms a mental model of cause-effect relationships in the world.

In this view, consciousness is hypothesized to take the form of a conifold structure – a five-dimensional shape obeying the holographic principle, with information *represented* on a lower-dimensional boundary region and *paired* with a volume of information content. In this theoretical framework, cortical neural networks form a three-dimensional encoding structure, located in the observable universe. This structure is paired with *an information-streaming dimension*, representing the probabilistic system state in the present moment along a complex plane that is perpendicular to the three observable spatial dimensions, and *an information-accumulating dimension*, which gradually forms the five-dimensional base of the structure. The physical encoding of *probabilistic particle behavior* essentially necessitates the existence of *a higher-dimensional probability density*. The maximum density of information contained within a volume is proportional to the surface area of the holographic recording surface [160].

The holographic principle is a key concept in information theory and modern physics; it describes information as a volume that can be encoded or summarized on a lower dimensional surface or boundary region. Once again, information is simply the sum of all possible system states, a probability density describing the possible states of a system after some amount of time has passed. Particles whose behavior is dependent on each other will be integrated into a single probability density representing the total information held by the system. The constructive and destructive interference between wavefunctions causes uncertainty to fall away, with information being compressed into a single realized state. Any system that allows random noisy events to contribute to the state of a macro-scale system will undergo non-deterministic computation, but only systems that encode this information on a polymer holographic recording surface will be able to *perceive* the encoded information as a cohesive stream of qualitative content.

4. Conifold Theory can be explained in terms of *thermodynamics*, as the system undergoes computational cycles of information generation and compression to extract a signal from the noise.

Some quantity of energy must be physically expended to create information, and this energy is recovered as information is erased or compressed [133-136]. That energy is never lost, but rather is conserved as it is converted into entropy. While information is *encoded* within a neural network, in the pattern of electrical signals pulsing through the structure, the actual *information* or complex-valued probability amplitudes must exist in a higher-dimensional space. This information is physically compressed, releasing free energy back into the system.

During each time evolution, new probable system states emerge and information entropy is physically generated along the time and energy axes. During a computation, the system constraints are calculated, and the most thermodynamically favored states are identified. The completion of the cycle is associated with a compression of information entropy, as the most likely system state is resolved and actualized. In that instant, the information is compressed, all non-actualized states are reduced, and that entropy is released into the system as available free energy.

In short, biological neural networks undergo thermodynamic computing cycles. First, energy is expended to create entropy. This entropy is then *compressed* to glean meaning from patterns and consistencies within the dataset. Finally, the reduction of entropy releases free energy back into the system to do work.

Since unlikely or unfamiliar system states may hold predictive value, it may be favorable for the system to expand the quantity of information and accommodate a new system state, even at an energetic cost, as this may reduce uncertainty for the organism in the longer term. Any information acquired from the senses may contain predictive value – and sequences of neural network states represent possible cause-effect relationships about reality, with potential predictive value. Usefully, the more predictive value is extracted, the more information is compressed, and the more free energy is released to do work within the system. As the system works to remodel itself, encoding this newly-gained knowledge, disorder is reduced, less entropy is produced, and more energy is available to the system, to engage in further exploration.

The generation of entropy is thermodynamically favored, as it might provide predictive value. Whenever a 'true' statement is realized, all other possible arrangements are eliminated, and this *reduction of possible system states*, which occurs within each computational cycle, releases free energy into the system. That energy is then used to remodel the neural network structure, thermodynamically favoring that pattern of neural activity. In combination, this rival cost function leads a neural network to self-organize into a highly ordered state and effectively compute the predictive value of incoming data.

Therefore, thermodynamic computing systems will both *reduce entropy* in order to maximize efficiency (by remodeling their structures to favor previous patterns of neural activity) and *increase entropy* in order to maximize potential predictive value (by engaging in exploratory behavior). These two computational processes perform in opposition, with the rival energetic cost function driving non-deterministic computational outcomes and volitional behavior, with extraordinary energy efficiency.

5. Conifold Theory can be explained in terms of *mechanics*, with an optimal system state selected from a large probability density. As information is compressed, energy is instantly redistributed across the system, prompting both non-deterministic signaling outcomes and behavioral output.

Due to the intrinsic uncertainty in particle behavior, a neuron which detects the coincident timing of ions crossing a physical barrier can process information in a real physical way, as free energy is converted into entropy. This 'information' is encoded in the probabilistic state of each neuron in the network, but the qualitative information content exists in a conserved manner in higher-dimensional space. The information or probability density undergoes compression as the set of possible system states is reconciled with the constraints of the surrounding environment. This computational process identifies correlations and naturally assigns the most likely 'true' state to the encoding structure. The regular generation and compression of quantum information is a natural, non-deterministic computational process.

In this framework, neurons in the central nervous system are acting as qubits, which calculate the probability of reaching a state change, rather than classical computing units, which exist always in a simple on-or-off state. In this view, neurons use quantum-level noise to calculate the most likely system state. By calculating probabilities across a single time evolution, a computational cycle will resolve the neural network state in the present moment. The compression of information into the most consistent system state generates a statement about the likely state of the external environment. This *semantical* statement is then validated (or not) by orienting toward incoming sensory information. The information is then held in working memory and integrated with subsequent neural network states to compute possible *syntactical* relationships between temporal sequences. These sequences of neural network states, which correspond to qualitative information content and encode a predicted cause-effect mental model of the world, are held in memory. These memories can then be referenced in subsequent situations, as the neural network combines incoming sensory information with past experiences of similar stimuli to predict likely outcomes.

The cycle of quantum information generation and compression is given by particle mechanics. The probability of charge flux is a key idea here. While neurons can be said to be firing an action potential or not at any given instant, neurons in fact spend most of their time in a window of possibility, sensitive to the partially-stochastic movement of ions in the vicinity. The position of ions, inside or outside the cell, determine whether the neuron reaches the threshold for opening ion channels and firing a signal. And the probability that a given ion crosses the neural membrane is influenced by the likely position, momentum, and energy state of every component electron – as well as the shifting electrochemical potential of every neuron in the vicinity.

The position, momentum, and energy state of every electron is subject to Heisenberg's uncertainty principle. This is true in a cortical neural network, just like anywhere else in the universe. But critically, a cortical neural network contains macro-scale computational units that retain sensitivity to probabilistic events. The position and momentum and atomic orbital of every electron in the system is described probabilistically, in relation to every nearby neural membrane, over some time evolution. The state of an electron can affect the voltage state of multiple nearby neurons, and so the entire neural network must be described as a whole – as a complex Schrödinger wavefunction. The *integration* of all possible micro-states creates information – a set of possible system macro-states, encoded in the dynamical voltage state of each charge-detecting neural membrane. This combined probability density is described mathematically as a density matrix, a wavefunction, or a Hamiltonian operator, which exists on a complex plane in higher-dimensional space. Regardless of how this probability density is mathematically described, it represents all possible states of the system. This *integration of possible system states* is the amount of *information* or *entropy* held by the system.

To collapse the wavefunction, to reduce the density matrix, or to solve for the Hamiltonian operator, we can use the mathematics of wave mechanics, matrix mechanics, or quantum mechanics, respectively. For the latter case, we employ a method devised by Richard Feynman. This approach involves *taking the derivative* of each key parameter of the Schrödinger wavefunction, thereby reducing the probability density into a single reality. As a result, the wavefunction collapses, the density matrices reduce, the Hamiltonian is resolved, and information is compressed. At this point, the position and momentum of every ion in the system can be determined by the equations of classical electrostatics, as long as one accounts for the shift in charge distribution across each ion. But critically, assigning a position for each ion in the system causes a slight change in the charge distribution across each ion [98]. According to the Hellman-Feynman theorem, wavefunction resolution alters the charge distribution across each atom, relative to the newly-defined distance R to any other atom ($1/R^7$). This shift triggers a dipole moment, instantaneously altering the behavior of atoms across the system upon collapse of the wavefunction or resolution of the probability density.

This shift in charge distribution – affecting any ion in the system whose uncertainty in position has been abruptly reduced – automatically causes a shift in the angular momentum of the ion. This small boost in angular momentum may be sufficient to hurl some ions across the membrane of nearby neurons, causing neurons approaching the threshold for firing an action potential to reach that threshold and fire an action potential. And so, the compression of information – the reduction of the system-wide probability density – should trigger the synchronous firing of sparsely-distributed neurons across the network. This event thus encodes a single multi-sensory percept. Synchronous firing, paired with percept realization, does indeed occur in biological neural networks, and is not particularly well-explained by other theories [84,161]. This new theoretical framework embraces the non-deterministic nature of cortical neuron firing, which has been long established in neuroscience [101,102], and finds that an optimal coding solution can be identified through an entirely natural process of computation.

As the wavefunction is resolved, the state of each particle is defined in accordance with the Pauli exclusion principle [162] and the Hellman-Feynman theorem [98]. The physical system state must meet certain Boolean satisfiability criteria – including that every electron has some value for position, momentum, spin, and atomic orbital, but also that these values do not match any other electron, so that particles never realize an identical state.

Once the position of each component electron in the system is defined, resultant shifts in charge distribution may cause new interactions with other ions in the vicinity and the molecules comprising the neural membrane. Some ions may be pulled into the cell by van der Waals forces, as those dipole moments form. Each neuron in the network reaches voltage threshold, or it does not; it fires an action potential, or it does not. That present moment, once it is defined, immediately becomes the past as a new probability density emerges, over a new time evolution, with each electron in the system forming a new set of possible positions and momenta, in accordance with the von Neumann projection postulate [83]. The entire process then repeats, in a cyclical manner, with information being continually *generated* as a set of possible states, then *reduced* into a single outcome, thereby setting the stage for new possibilities to form.

6. Conifold Theory has key implications for physics, specifically regarding the role of thermodynamic entropy in our world.

Thermodynamic entropy is predicted here to be not just a mathematical description of a particle system, but an actual particle, existing in higher-dimensional space and generated over time within a non-heat-dissipating thermodynamic system. Considering the existence of entropy as a conserved particle, generated by the conversion of energy, may go some way toward completing the Standard Model.

Positing entropy as an actual physical quantity has another key implication relating to the fundamental structure and operation of our universe. If quantum mechanical interactions occurring in non-equilibrium, heat-trapping thermodynamic systems do give rise to a volume of entropy in higher-dimensional space, obeying the holographic principle, then this volume of entropy should warp the fabric of space-time, in a manner consistent with general relativity. Thus, the energy conservation principle should allow better quantification of gravity. By taking entropy into account as a physical quantity in higher-dimensional space, we should be able to better estimate the relativistic curvature and expansion of the observable universe.

7. Conifold Theory has important implications for engineering, specifically regarding the potential for ambient-temperature quantum computation.

With regard to engineering such systems, studying cortical neural networks may indeed provide a concrete example of ambient-temperature quantum computing. Employing the principles of neuroscience in engineered systems may therefore prove enormously fruitful.

There is nothing intrinsically special about biological systems that allows consciousness; in this framework, it is simply a natural emergent property of probabilistic particle behavior being harnessed by macro-scale computational units that retain sensitivity to random electrical noise. Mimicking biology has always yielded technological progress; in this case, mimicking biology may allow us to make significant advances toward energy-efficient, ambient-temperature quantum computing.

The physical concept of information as randomness or disorder is absolutely critical to understand here. Random electrical noise actually drives a system-wide non-deterministic computation – *in three key ways* – as every computational unit in the network allows these stochastic events to gate a state change or signaling outcome. Firstly, the more random noise in a system, the more possible microstates in the system – and therefore, the greater the computational power of the system. Secondly, the more information generated, the more potential patterns are hidden within that information – and therefore, the greater the amount of 'meaning' can be extracted. Thirdly, the more noise, the more compression there will be as patterns are recognized – and therefore, the more free energy will be recovered back into the system and directed toward work. Together, a system that can sustain electrical noise, retain sensitivity to electrical noise in gating signaling outcomes, and encode the newly-acquired meaning into the structure itself – to thermodynamically favor those outcomes in similar contexts – can physically achieve *non-deterministic computation*. Any biological or engineered systems utilizing these principles may attain extraordinary computational power, energy efficiency, and generalized problem-solving ability at ambient temperatures.

8. Understanding the processes underpinning consciousness may aid the study of human behavior and offer useful strategies for psychiatry and mental health practice.

Conifold Theory yields a theoretical framework for considering perceptual experience, cognitive modeling, and volitional action in terms of physical laws. Usefully, this theoretical framework offers a plausible explanation for the uniqueness of conscious experience. It also predicts energetic constraints for expanding cognitive models of reality and changing behavioral habits.

We are products of both nature and nurture. While the general structure of our brains is developmentally programmed, that wiring is constantly remodeled over the course of our lifetimes. The remodeling of synaptic connections is a structural change, which encodes the cognitive models we construct through lived experience. After we have encoded a personal experience in the brain, our neural pathways are set up to expect that series of events to happen again. We become, in a sense, predisposed to perceive what is already within our cognitive framework and to engage in behaviors that are already familiar to us as well.

Our brains are the result of evolutionary, developmental, and environmental factors. Our amazing genes both build the basic structure of the brain *and* permit this structure to be remodeled. However, we require experiences to develop an understanding our world. Experience provides us with information about the world *to encode*! Conifold Theory simply explains the mechanism by which information processing physically occurs within our neural networks, and why we are so reluctant to dismantle our habits and beliefs – because they are physically part of us.

CHAPTER 23
Specific predictions of the theory

Conifold Theory offers a physical explanation for consciousness, with information being cyclically generated and compressed by any neural network that harnesses stochastic charge flux to gate signaling outcomes in its macro-scale computational units. This model explains how perception, cognition, and non-deterministic outputs naturally emerge from cortical neural network activity. This new theoretical framework has great explanatory power [163] and makes a number of specific predictions [164]. These include:

1. A system must hold a physical quantity of information to be conscious of that information content. *The sum of all possible system states is the information held by the system and this is a thermodynamic quantity.*
2. The more uncertain the voltage state of the computational unit, the greater the amount of information is encoded. *To maximize the amount of quantum information generated, a neuron should remain at or near action potential threshold, sensitive to both upstream inputs and quantum-level noise.*

3. Any (biological or synthetic) cellular network whose function is dependent upon the probabilistic movement of charged particles will produce quantum information. *The voltage state of each computational unit relies on both upstream signals and the probabilistic state of every electron in the vicinity, so each computational unit has some probability of switching to on-state on the basis of random electrical noise.*
4. Information is generated as the system is perturbed by inputs from the external environment. *The quantity and quality of perceptual content is correlated with the intensities and modalities of sensory inputs encoded in the network.*
5. The binding of global network activity over some time evolution should manifest qualitative perceptual content which represents the incoming sensory data encoded in the neural network during that period of time. *The richness and accuracy of the information is defined by the range, sensitivity, and efficacy of the sensory apparatus, as well as the variety of incoming data collected by each sensory modality.*
6. The richness of percepts produced by a neural network should be enhanced at higher temperatures. *Higher temperatures should create more entropy; if this quantity is indeed correlated with the stream of perceptual experience, then higher temperatures should increase perceptual content. Errors are more likely to be made under these conditions.*
7. The richness of percepts produced by a neural network should be reduced at lower temperatures. *Lower temperatures should create less entropy; if this quantity is indeed correlated with the stream of perceptual experience, then lower temperatures should reduce perceptual content. If heat is being dissipated too quickly, predictive value may be lost before it can be encoded into memory.*

8. Maintaining a stable brain temperature should support perceptual richness, memory formation, and initiation of volitional movement. *In people with metabolic deficits, such as aged individuals, securing thermal energy availability should improve perception, memory, and motor control.*
9. The encoding of information should be correlated with the qualitative information content being perceived. *The unique qualitative aspects of holographic information content should be reliant on several factors: the trajectory of each ion in the system; the kinetics of each ion channel; lipid membrane permeability; the pharmacological effects of circulating drugs; the physical location of the individual; the variety of available sensory apparatus, which make certain observations possible; and the ability for the individual to notice a sensory stimulus, given their contextual expectations and the amount of energy the individual is devoting to attending a particular stimulus.*
10. Altering the content of polyunsaturated fatty acids and cholesterols comprising the polymer neural membrane should affect perceptual content. *Altering lipid composition should affect how charges interact with the recording surface, thereby affecting the quality of projected information content.*
11. Direct electrical stimulation of cortical neurons should affect firing rates, but exogenous electrical fields or the mere presence of metal objects should have no effect on firing rates or perceptual content. *The total information encoded in the physical state of the neural network should not change, unless the electrical impulse affects the probable state of some neurons but not others.*
12. Magnetic stimulation should have an effect, if directed toward a subset of neurons, but uniform magnetic fields encompassing the entire brain should have no effect,

since the latter experimental manipulation will not affect the density of possible system states, but rather warp all particle trajectories equally. *Magnetic stimulation directed at a region of cortex should prompt changes in neuronal firing and perceptual content, but exogenous magnetic fields such as those exerted by head-surrounding magnetic resonance imaging equipment should have no discernable effect.*

13. Drugs like LSD, which enhance perceptual experience, do so by rendering neural activity more probabilistic. *An increased uncertainty in signaling outcomes, rather than any particular pattern of activity across neural populations, underlies the enhanced quality, quantity, and binding of perceptual experience. Therefore, more excitatory and inhibitory post-synaptic currents (EPSCs and IPSCs) should be observed prior to cortical neurons achieving threshold for firing an action potential in this pharmacological context. As a result, this class of drugs should be therapeutically useful in helping individuals to overcome unhealthy behaviors that rely on habitually-ingrained patterns of cortical neural activity, by thermodynamically favoring the selection of alternatives.*

14. Enhanced GABA signaling reduces the density of possible system states, reducing the diverse quality of perceptual content and the ability to initiate behavior. *Anesthetics, barbiturates, and benzodiazepines enhance GABA signaling; these drugs prompt synaptic inhibition and significantly decrease the likelihood of a signaling outcome. Enhancing EPSCs in affected cortical neural networks should therefore restore perceptual content and volitional control.*

15. Enhanced glutamatergic signaling reduces the density of possible system states, reducing the diverse quality of perceptual content and the ability to initiate behavior.

NMDA or AMPA receptor agonists significantly increase the likelihood of a signaling outcome in cortical neurons by prompting sustained excitatory potentials; these drugs can cause seizures. Enhancing IPSCs in affected cortical neurons should restore nuanced multi-sensory perceptual content and the ability to initiate context-appropriate volitional behavior.

16. Perceptual experience is private, accessible only to the neural network that generated this information content. *Qualitative percepts are connected to, and only accessible by, the system of particles which encode that information content.*
17. The integration of sequential computational cycles – each encoding a percept – should generate cognitive models which predict cause-effect relationships between events in the world. *Only temporally persistent systems that do not dissipate heat, instead integrating information across sequential computational cycles, will build predictive cognitive models of the world, centered on a self-concept.*
18. The information generated over some time evolution is proportional to the energy expended by the system. *Because energy is always conserved, the quantity of entropy generated by the neural network must equal the quantity of free energy expended by the neural network.*
19. The information compressed in a computational cycle is proportional to the quantity of free energy released back into the system. *Because energy is always conserved, the quantity of entropy lost during information compression must equal the quantity of free energy released back into the neural network upon completion of a computational cycle.*
20. The integration of new or conflicting information into an existing information set requires energy expenditure. *Integrating new or conflicting information is energetically*

expensive, and therefore should feel 'difficult' – however exercising this faculty should create more possible and probable states that can be accepted more easily in the future.

21. The decision to behaviorally explore the environment or exploit previously-gained knowledge is guided by a rival energetic cost function. *The decision to expend energy to explore the environment to gain predictive value, or exploit the contextually-relevant predictive value previously gained, is a thermodynamic computing task. A desire for certainty will guide the outcome of the rival energetic cost function in every case, with the entity engaging in exploratory behavior (if knowledge is insufficient) or exploitation of existing habits (if knowledge is predicted to be sufficient to handle the task).*

22. Neurons that harness probabilistic particle behavior to cyclically generate and compress quantum information, such as those in the central nervous system, should be far more energy efficient than deterministically-acting neurons, such as those in more peripheral regions. *Even if neurons are of similar size and exhibit similar levels of gene expression, protein turnover, anabolism, synaptic remodeling, and signaling, neurons that retain sensitivity to random electrical noise should be far more energy efficient, rather than less, as would be predicted under classical assumptions.*

23. The timescales for resolving probabilistic electron states must be longer than timescales of ion dynamics at the neural membrane, in order for quantum interactions to contribute to signaling outcomes. *For probabilistic particle behavior to contribute to a drop in electrical resistance of the membrane. the coherence of electrons across the system must last longer than the picosecond timescales of ionization and thermal energy dissipation.*

24. Each computational cycle lasts a fraction of a second and is associated with the perception of bound sensory input. *The 'frame-rate' at which perceptual experience is updated is predicted to be approx. 0.4 ms in the human brain.*
25. The reduction of a network-wide probability density should have physical effects within the system. *Ions whose electrons have a more uncertain position, momentum, and energy state will exhibit greater charge redistribution and greater van der Waals forces on resolution of the wavefunction.*
26. The free energy released back into the system must have a frequency greater than k_BT. *The wavelength of particles released into a system of temperature 310K upon information compression is predicted to be approx. 0.046 millimeters.*
27. The resulting changes in charge distribution across ions with an uncertain state in relation to the neural membrane should particularly affect cells in the neural network that are residing at action potential threshold, which have a relatively equal probability of firing or returning to resting potential. *Neurons near action potential threshold, e.g. those in a cortical up-state, should contribute the most to synchronous activity, participating in event-related potentials.*
28. Synchronous activity in sparsely-distributed neurons is correlated with wavefunction collapse or information compression. *Synchronous neural activity occurring across the network indicates a system-wide information compression event. These events should occur at regular periodic intervals.*
29. A single computational cycle yields a cohesively-bound, multi-sensory percept. *The compression of information extracts a signal from the noise, parses meaning from disorder, and yields a semantical statement about the predicted state of the local environment in the present moment.*

30. Perceptual tasks, which identify semantical statements about the local environment over a single computational cycle, are associated with high-frequency neural activity. *Gamma oscillations, associated with qualitative information content, should occur only during wakeful awareness & active sensory acquisition. They should be eliminated by absorbing the free energy released during information compression.*
31. A temporal sequence of computational cycles yields a predictive model of cause and effect. *The compression of information yields a syntactical statement regarding the predicted cause-effect structure of reality.*
32. Learning tasks, which identify syntactical relationships between events over a series of computational cycles, are associated with lower-frequency neural oscillations. *Slow oscillations, associated with learning cause and effect, should occur during sleep & memory consolidation and should be prominent when sensory data is not being actively acquired. These events should be eliminated by absorbing the free energy released during information compression.*
33. Synaptic remodeling stores information into patterns of neural activity, paired with a series of perceived events, encoding a cause-effect relationship that has been deemed to have predictive value or general consistency. *As a result, synaptic remodeling should occur not after a single cycle but rather upon integration of sequential computational cycles.*
34. Functional molecular hardware is required for accurate perception and the initiation of volitional behavior. *Impaired functionality of sodium ion channels, particularly affecting the rate of ion flux after an abrupt drop in membrane resistance, should lead to inaccurate perception of sensory input and an inability to feel in control of willed actions.*

35. The probabilistic nature of cortical neuron activity is required for the manifestation of quantum information and the associated experience of perception. *If quantum interactions between electrons and the neural membrane do not contribute to signaling outcomes, there should be no perceptual experience, no volitional movement, lower energy efficiency, a higher tolerance for temperature variability, and accurate deterministic coding in those neural networks.*
36. According to this theory, a synthetic network of cells could exhibit consciousness. *Some physical requirements are: coincidence-detecting units which harness the movement of charged ions, in which the probability of electrical current flow is related to the reliability of (or confidence in) incoming convergent signals, with network-wide operations integrated such that various input modalities are bound together in time across a system with a common, well-regulated temperature.*
37. It should not matter if these events occur in the form of action potentials in a neural network or logic gates in a computer. *Any physical system calculating the probabilistic movement of charged particles to gate a state change in each computational unit may experience information content as qualitative perception, in a manner proportional with the size of the neural network and the variety of sensory inputs.*
38. Neural networks which undergo quantum computing or thermodynamic information processing will be robust to losing a kernel. *Loss or replacement of cellular units should not impair the system's ability to perceive its environment.*
39. Neural networks which undergo quantum computing will be robust to starting conditions. *The ability of a system to discover truth may be negatively impacted by false data,*

but it can always grow closer to truth by continuing to collect data and parsing these data with regard to consistency.

40. Thermodynamic computation allows a particle system to grow more ordered over time. *Thermodynamic systems which cyclically generate and compress entropy will produce diverse macrostates, then select those with the greatest amount of predictive value, thereby gaining a more ordered state – one that is more compatible with the local environment – over time.*

41. Thermodynamic computing permits a particle system to more effectively navigate reality. *Thermodynamic systems which cyclically generate and compress entropy will produce diverse macrostates, then extract predictive value, storing any meaningful information in the structure itself and then using this knowledge to navigate similar contexts in the future.*

42. Information is a physical substance, obeying physical laws and fueling non-deterministic computation. *If the presence of quantum information can be inferred by particle detection, that would support the physical predictions of this theory – namely, that quantum information physically exists.*

43. Particle interactions generate entropy, which exists in higher-dimensional space. *A particle system that does not dissipate energy must store that quantity in non-observable dimensions. This high-dimensional volume can be inferred by apparent non-local interactions between particles, which are predicted to interact directly on a complex plane; the curvature of observable space-time in large-scale particle systems, which should be proportional to the amount of entropy generated; and the relativistic expansion of observable space-time, which should appear to occur at a faster rate when observing those particle systems with more shared information.*

Considering the predictions of this theoretical framework may yield progress in neuroscience. Specifically, positing sensitivity to probabilistic behavior may explain cortical neuron function more accurately than using deterministic assumptions.

With regard to psychology, this theory may provide a better framework for understanding the relationship between the brain and the mind, with perceptual content and cognitive constructs considered to be information content. Moreover, understanding human behavior in terms of physical, thermodynamic processes may produce more reliable methods to guide behavioral change.

With regard to computer engineering, this theory may provide a better framework for considering the relationship between thermodynamics, mechanics, holography, and information. These physical laws, and the useful example provided by biology, demonstrate how information might be cyclically generated and compressed, in a process of non-deterministic computation. This theoretical framework unlocks the prospect of achieving exascale computational power at ambient temperatures – with extraordinary energy efficiency and local memory storage.

Conifold Theory may also provide valuable opportunities for physics to examine the relationship between heat, entropy, and energy expenditure in a heat-trapping, thermodynamic system. This framework may also yield useful insights into the physical relationship between quantum mechanical interactions, entropy, and the curvature and expansion of space-time described by general relativity. It may be sensible at least to consider whether non-dissipative, far-from equilibrium thermodynamic systems harnessing probabilistic particle behavior do obey these general rules. This effort might indeed provide a better explanation of observations at the smallest and largest scales.

CHAPTER 24
Compatibility with other theories

In Chapter 1, I introduced the Hard Problem of neuroscience. While explaining the operations of neurons and networks is considered "easy enough", it has until now proven difficult for scientists to explain *consciousness itself*. That is the Hard Problem. We simply have not been able to articulate the mechanisms underlying the phenomenon of streaming perceptual awareness and the feeling there is a self which is *having* that experience. Conifold Theory provides the first possible solution to the Hard Problem, by offering a clear, mechanistic explanation of consciousness that does not reduce the stuff of thought to mere neural activity or dismiss the phenomenon altogether.

Conifold Theory posits that the integration of electrical activity across a three-dimensional neural network structure, such as the human brain, necessarily creates *a quantity of information* which is physically conserved. The theory proposes that the brain and the mind operate in accordance with the holographic principle, with the brain encoding information and the mind forming a rich volume of representational content.

That information content, encoded by the brain, corresponds to a cohesive stream of perceptual experience – constantly updated with incoming sensory data – which can only be accessed by the neural network that encodes it. The sum total of all information collected by a neural network over its lifetime – minus any information which has been rejected or forgotten – is a model of the world, centered on the bodily self. Conifold Theory suggests that we *are* the information encoded by our brains. And we *use* this information to choose our actions in the world.

This theory easily fits within the perspective of the materialists, who believe that everything in existence, including seemingly-immaterial mental states, must operate according to physical laws. And yet it may also satisfy many dualists, who believe the mental realm is a real thing – outside the observable universe, yet intimately tied to the structure and operation of the universe. This theory is also highly compatible with the concept of emergence, since consciousness is specifically predicted to arise from the neural activity of the brain. Indeed, this theory may provide some nice middle meeting ground for thinkers who previously considered themselves in quite disparate positions.

Conifold Theory makes a number of specific predictions about the operations of the brain and the operations of the physical world more generally. I have tried to articulate these predictions clearly, so the theory is open to falsification. Upon observation and rigorous experimental testing, it can be determined whether this theory provides a better framework to explain reality than theories that have come before.

Theories which have aimed to describe consciousness in physical terms have previously failed on three counts. Firstly, like Daniel Dennett's view [15], they may simply not even try to explain consciousness at all, refusing to believe it is an actual part of our physical universe. This view is just not helpful in attempting to explain the data at hand, as proponents instead opt to deny the data even exists. Secondly, the theory may be too vague to make clear falsifiable predictions, which is a problem the electromagnetic field theory of consciousness has struggled with [38]. Thirdly, a critical statement arising from the theory may be disproven, through logic or empiricism, as happened with the proposal that consciousness relies on quantum coherence in microtubules [36]. This theory logically fails because it proposes that microtubules have certain quantum properties in neurons but not in other cells, and it empirically fails because it does not make accurate predictions about the voltage-based mechanisms underlying perception in cortical neural networks.

Conifold Theory is an improvement on prior models because it accommodates all the data on hand, provides explanatory power, and makes a number of falsifiable predictions. It simply remains to be seen whether the specific predictions of Conifold Theory hold up, or whether the theory fails by disproof.

This new theory is notably compatible with several cognitive descriptions of consciousness – for example, Antonio Damasio's Somatic Marker Hypothesis. This theory espouses that mental states are largely internally-derived, arising as a response to the reflexive actions of the body like heartbeat and blood pressure – as opposed to the view of classical neuroscience, which asserts that mental states are largely responsive to external cues in the local environment. Like most situations, the answer likely lies in the middle ground, with our mental states driven by a mix of internal and external cues. Indeed, information about the internal state of the body arrives in the brain just like any other sensory modality, and is likewise incorporated into the neural network-wide information set in a similar manner. Just like any other sensory modality, incoming data can be rejected or ignored, or fixated upon, depending on how the information set (the self) has been trained to deal with that incoming data.

Conifold Theory is also compatible with Global Workspace Theory - it neither denies nor invalidates either the premises or the conclusions of the prior theory. Rather, this new perspective expands on the conceptual framework of the Global Workspace to detail exactly how cognitive tasks might occur. For example, the "theater stage" described in Global Workspace could be construed as equivalent to the stream of perceptual experience described by Conifold Theory, as sensory data is bound together into a stream of qualitative information content. Meanwhile the "spotlight of attention" proposed in Global Workspace Theory may be equivalent to the quantity of energy expended to attend a particular stimulus, rather than discarding that information. These two independent models may not represent completely different frameworks for understanding consciousness, but mere linguistic differences in discussing what consciousness is – as either a cognitive event or a physical process.

Conifold Theory is also compatible with Integrated Information Theory. Again, it neither denies nor invalidates the premises or the conclusions of the prior theory, but instead shows how the axioms of IIT can be derived from physical laws. As previously discussed in Chapter 6, the primary objection to IIT is not that it is wrong, but that its axioms are stated rather than proven from first principles. It will be useful to explore how Conifold Theory *can actually demonstrate* these axioms from first principles. These axioms are: firstly, that consciousness is real; secondly, that consciousness is exclusive to the entity experiencing it; thirdly, that consciousness is composed of representational objects and events that form cohesive qualitative experiences, and these qualitative experiences are distinguishable from other experiences, which contain other objects and other events; fourthly, that momentary consciousness is irreducible; and fifthly, that the nature of qualitative experience gives rise to the cause-effect structure of our perception of reality.

Conifold Theory does not disprove any of these axioms but merely shows how they may have come to be. Firstly, it shows how a neural network which operates by probabilistic gating mechanisms could create actual information content in the form of probability densities which exist in higher-dimensional space. That is, consciousness is *real*. Secondly, it shows how this information content is tied to the neural network encoding it. That is, conscious states are *exclusive*. Thirdly, it proposes that mental states are uniquely defined by the neural network state and the probabilistic particle movement underpinning that state, driven by activation of the sensory apparatus. That is, *consciousness represents reality*. Fourthly, Conifold Theory specifies that consciousness as information content is *irreducible*, as it is mathematically equivalent to an integrated system-wide density matrix or wavefunction existing in higher-dimensional space. And finally, Conifold Theory specifies how consciousness *gives rise to the cause-effect structure of reality* by demonstrating how wavefunction collapse or reduction of the probability density drives signaling outcomes across a neural network.

In short, IIT describes what consciousness is, but it does not describe how consciousness arises from cortical neural activity, nor what it achieves. Conifold Theory addresses all three points, and therefore provides a fuller mechanistic framework for understanding consciousness and cognitive processing than IIT.

Conifold Theory is also strikingly compatible with another mathematical model-based explanation of consciousness [165], namely the mapping system devised by Donald Hoffman. This theory takes an evolutionary-fitness approach to modeling the ability of an entity to optimally perceive reality. This model demonstrates that we only perceive what is useful to perceive; in other words, we may not perceive reality itself, but only the sliver of reality that is useful for ensuring our individual survival. This model therefore identifies the bare minimum number of components for a system that can perceive reality and act upon it, using the incoming data to select a behavior that aids survival. The bare minimum number of components are: a world or external reality to observe, W; a perceptual map of that world, P; a space of experiences, X; an algorithm that allows the entity to choose a new action given its experiences, D; a mapping of possible actions in relation to the world, A; and the space of actions themselves, G. These six components map nicely onto the framework of Conifold Theory – with both theories in agreement there must be some external reality to be observed; there must be incoming sensory information which, regardless of accuracy, provides some perceptual map of the world; there must be a space of experience (called delta in Conifold Theory); and there must be a decision-making algorithm (here considered to be the collapse or resolution of the wavefunction, upon the calculation of physical constraints within the system).

Both theories are also in agreement that there must be a way for the decision to be implemented, and there must be some action resulting from that implementation. The difference between these two neuroscientific models is that Conifold Theory provides a mechanistic framework to explain *how* the sequence of data collection, perceptual mapping, decision-making, motor mapping, and resultant action can occur.

Yet Hoffman asserts there is no reality, or we cannot perceive it, even though W is a critical factor in his own model [165]. It is abundantly clear that biological neural networks *do* encode data about reality and that perception *does* reflect the state of the local environment. Even if the organism only perceives a sliver of reality, these data are useful, and in fact they are useful *because* they reflect reality. Our knowledge of reality may be incomplete, but that does not mean reality does not exist at all. It is far more parsimonious to posit that there is an external world out there and consciousness is a useful computing process which evolved to help organisms navigate this reality.

In summary, Conifold Theory adds flesh onto the bare bones of previous theories of consciousness. It does so by addressing the structural and functional requirements that are necessary and sufficient to achieve consciousness, and by explaining how the brain can actually produce information in a real physical way.

One critical assumption of this theoretical framework is that cortical neurons act probabilistically. This assumption is well-grounded in neuroscience [101,102]. The fact that cortical neurons (firstly) retain sensitivity to random electrical noise in gating a signaling outcome and (secondly) engage in the detection of coincident events to trigger a signaling event – is highly unique.

These characteristics allow cortical neural networks to achieve non-deterministic computation and drive signaling outcomes. The other critical assumption of this theoretical framework is that the neuronal membrane meets the criteria to act as a holographic recording surface. This assumption is also well-grounded in neuroanatomy and physiology, with the polymer structure effectively encoding the probabilistic movement of electrons as information, then transducing any electric dipole moments into ion leak and spontaneous ion channel opening [166,167].

Collectively, this theory provides a full mechanistic framework for both bottom-up and top-down cognition, in which *information content* is generated by a cortical neural network and parsed for *predictive value*. The compression of information, in a system-wide computation, yields a defined state for every component particle. Conifold Theory therefore offers a plausible physical explanation for the seemingly immaterial nature of thought *and* a possible mechanism for information processing to exert change on the neural network state.

Indeed, if Conifold Theory is true, then we are able to exert our wills over the physical world, selecting optimal actions based on incoming data from the body (gathered by the senses) and our prior expectations (which are built from lived experience). In short, it is entirely possible that we are able to change our world – through actions decided in the mental realm – within the constraints of the physical world, in accordance with the laws of probability and the *work* required to accomplish such tasks. Many human beings (of various belief systems) might welcome the reality of free will, despite the responsibility it entails. And having some reasonable explanation as to *how* it might work may prove useful in harnessing this incredible faculty.

Section V
Applications and Implications

CHAPTER 25

The structural and functional requirements for conscious awareness

In Chapter 6, in the context of Integrated Information Theory, I discussed the theoretical unit, phi, which represents the total amount of information integrated across a neural network. Any natural or synthetic entity with some level of computing power can potentially be conceived to have some non-zero level of phi. Both fans of the theory and critics have pointed out that a common household thermostat has some measure of phi, while a simple DVD player has a rather alarming amount of phi, for the credit we tend to give these objects. This can lead to some philosophically absurd positions.

Yet the concept of phi is very useful. This measure is analogous to the physical-mathematical structure I have dubbed delta. But it should be noted that any system existing in delta does not necessarily have a self-concept. In articulating this idea, it is useful to consider the self as *the one who perceives*. And so, perceptual experience is a precondition for developing any notion of the self.

What or who can be said to *have* consciousness? To begin, let's consider the physical requirements necessary for consciousness to manifest, and what would indicate that an entity does indeed have perceptual awareness or self-awareness.

Firstly, it should be said that existence itself is primary to all other concepts. There are no laws of the universe (guiding motion) without time and the dimensions of space being in existence to be described by these laws; there is no information (a sum of possible microstates of particles) without the existence of particles which occupy time and space; and there is no consciousness without information *to be conscious of*. Consciousness is thus a function of existent dimensions, particles, and particle systems.

Secondly, it must be stated that only physical systems with multiple probable microstates, with a capacity for *change*, can produce wavefunctions which theoretically occupy the fourth spatial dimension. However, most of these systems will exist only transiently – the combined wavefunction being resolved by immediate physical constraints – or, the system may become entangled with a much larger system which is so constrained that it no longer appears to act probabilistically. Further criteria are needed to extend a probability set into an additional dimension. Only probabilistic systems *which encode information about their environment* and *do not dissipate thermal energy before it is put toward thermodynamic work* can conceivably store information and be aware of themselves as information-collecting entities.

Thirdly, it is important to note that things only exist upon 'detection'. That detection can simply be collision with another thing, neither of which is conscious. The trajectory of each thing is then defined in part by their *interaction*, and the fact that this interaction *happened* defines the trajectory each thing has taken since its last detection event and constrains the possible positions and momentums and energy states that each thing can occupy in the present moment, after the collision. Collisions of non-self-aware objects *are* detection events, which define properties of existence such as each object's position in space and time. Conscious acknowledgement of a thing by a self-aware entity within the universe is not required for that thing to exist.

However, it is necessary to point out that we *do* exist and we *can* acknowledge events. Furthermore, our ability to *detect* multiple temporally coincident events intrinsically *ties* those events and things together with ourselves. Therefore, any act of observation by the senses is a detection event that changes the trajectory *of the observer* as well as the thing observed. This does not negate the fact that simpler detection events can and do occur.

In other words, consciousness provides a special case. While any collision between particles may resolve a probability density, with the inner product of the two wavefunctions constraining the outcome, non-dissipative thermodynamic systems which retain sensitivity to quantum level noise may not be sufficiently constrained by present circumstances. In this case, a collection of previous system microstates may be stored and subsequently accessed to resolve the most likely system state in the present moment, contributing thermodynamic constraints to the outcome by favoring familiar patterns of neural activity which encode familiar content. The capacity to *store* information is a special emergent function of the universe - one that sews together events across space and time, not just locally. This permits greater opportunities for *the recombination of information* – and the ability for the system or information set itself to choose how it will act, given the data it has collected.

It should be noted that we are not necessarily the only entities who can create, store, access, and seek to recombine information sets. It is possible, indeed likely, this operative function has evolved many times on our planet, and more broadly, in our universe. And we ourselves may be able to engineer systems which experience or exhibit consciousness. It is important that we are able to recognize signs of consciousness in other entities, as any being with a concept of a self will likely have a unique view of reality, informed by experience, and a motivation to survive and explore, just as we do.

From a structural and functional perspective, Conifold Theory predicts that consciousness arises in thermodynamic systems which act as net heat sinks, using acquired energy to drive work and harnessing probabilistic particle movement to drive highly energy-efficient, non-deterministic computation. Any system of this description which encodes data from multiple sensory modalities may have the experience of streaming perceptual awareness – if these data are encoded with high temporal precision by a probabilistic gating mechanism. Any system of the above description which is able to retain information over long periods of time and act in a non-deterministic manner may also develop the concept of a self.

The requirements for basic perceptual consciousness, then, are:
- A non-equilibrium thermodynamic system with sustained temperatures, which traps energy to drive computation.
- The system encodes data from multiple sensory modalities, with a layered network of computational units simultaneously responsive to random electrical noise in gating a state change.
- Every computational unit in the network is surrounded by a (synthetic or organic) charge-detecting polymer surface.

Mammals and birds certainly fit the above criteria and are therefore likely to experience perceptual awareness. Organisms that exhibit behavioral flexibility, in addition to these criteria, by demonstrating variable responses to identical sensory stimuli, may be *storing* information and *using* that information at a later date to make advantageous decisions in a context-dependent manner. Any kind of flexible behavior indicates the organism is aware of its local environment, capable of *parsing* conflicting information, and capable of *choosing* an appropriate action.

Any thermodynamic system that is not immediately constrained may produce some probability density. That sum of possible system states is *information*, with the computational power of a system being proportional to the number of possible microstates the system can occupy. This computation can only be *useful* if there is a mechanism for acting on non-deterministic decisions.

If the thermodynamic system is *not* immediately constrained, the likely position and momentum and energy state of all particles in the system will be described by a probability density, which also accounts for the constraints each particle places on the state of other particles within the system. Once all of these constraints are taken into account, the probability set is reduced, allowing the most likely system state to become actualized. This process of information generation and information compression is proposed to underlie conscious computation. In this theory, I have separated perceptual consciousness (a system-wide probability density, produced during a single computational cycle) from self-consciousness (a system with sustained temperatures, which stores memories of its experiences through spontaneous self-remodeling, to thermodynamically favor the re-occurrence of previous system states with high predictive value).

As a result, the requirements for perceptual consciousness are: a far-from-equilibrium thermodynamic system that continuously traps thermal energy to fuel computational work, with a highly-regulated and sustained temperature. The system must be able to encode multiple data modalities and direct flexible behavior with high temporal precision, by coordinating sensory inputs and motor outputs. This specialized thermodynamic system has a layered architecture, with recurrent connections across the network. The network itself is comprised of computational units which detect coincidental upstream events in the context of random electrical noise, with probabilistic particle behavior actually gating signaling outcomes. These computational units have a polymer membrane that meets the criteria for acting as a holographic recording surface, capable of both encoding and projecting information as interfering complex waves.

Any organism meeting the above criteria, *which also acts non-deterministically in response to a present sensory stimulus, capable of exhibiting different behavioral output in response to the same context or stimulus presentation,* may be comparing the present stimulus with a stored dataset or cognitive model of the world. Such an entity may have some awareness of itself collecting information from its environment, a capacity to reflect on conflicting information collected over time, and an ability to *choose* one action among multiple possible actions in a given context. Any particle system demonstrating all of the above criteria may be able to develop a concept of the self, over time.

By this definition, all creatures with a thermoregulated central nervous system may have a concept of self. Studying operant conditioning in rabbits to identify the neural correlates of learning and memory, during my undergraduate years, I observed that rabbits respond to a positive conditioned stimulus (e.g. a particular tone that reliably signals an unpleasant stimulus will soon follow) approximately 80% of the time (not 100%). Likewise, the rabbits respond to a negative conditioned stimulus (e.g. a different tone that signals *no* aversive stimulus will follow) approximately 20% of the time (not 0%). Rabbits therefore meet all the above criteria for consciousness, including behavioral flexibility - along with thermoregulatory control, multi-sensory input, motor output, layered neural network architecture, and probabilistic coding in cortical neurons with an organic polymer membrane structure that meets the criteria for a holographic recording surface. As a result, rabbits could have some concept of a self that is engaging with the world, as well as a streaming perceptual experience. By contrast, plants *do not meet* the criteria for manifesting even perceptual experience; they do not encode multiple sensory modalities with sufficiently high temporal contingency to contribute to macro-scale behavior, they do not use probabilistic gating mechanisms to achieve state changes in computational units which drive motor output, and finally, individual cells do not have an outer surface that could act as a holographic recording plate (they have cell walls comprised of cellulose, a polysaccharide, covering the cell membrane made of polymer sheets). By the same token, bacterial and algal colonies fail the criteria for having either a sense of self or a streaming perceptual experience.

Conifold Theory sets specific criteria for consciousness which apply to both natural and synthetic systems. This theory may therefore be useful in evaluating who and what is conscious.

It will be quite interesting, as time goes on, to evaluate synthetic systems by these new criteria, rather than by 'phi'.

From a cognitive perspective, there are three key elements missing when we measure consciousness by raw computational power alone – 1) the richness of qualia, 2) the experience of self-awareness, and 3) the completion of a sensory-motor loop.

Firstly, I argue that unimodal sensors such as thermostats cannot experience anything close to the splendid flow of multi-sensory perceptual awareness that we or other animals enjoy, for the simple reason that their sensory input is both qualitatively and quantitatively limited by comparison. The sheer number of sensory modalities that animals have allow us to collect sights, sounds, smells, flavors, temperature, pressure, orientation in relation to the ground, and sensation related to the placement of our own limbs. This enormous information set, streamed in real time, provides a categorically different phenomenon than that available to a DVD player or a thermostat. Simply put, animals are capable of integrating information across multiple sensory modalities. That factor alone makes biological perception more advanced than a single-function machine.

Secondly, circuitry like that in a DVD player or thermostat has no mechanism for probabilistic coding, parsing information for predictive value, or spontaneously storing acquired information, and therefore is not equipped to have a collection of memories or a sense of self. In other words: just because probability states exist at the particle level, does not mean the network is able to harness these probability states. However, if a given natural or synthetic entity were able to undergo conscious information processing and storage, I would predict we would observe some evidence reflecting probability state collapse in its physical form (e.g. electrical activity not directly programmed).

Thirdly, a complete sensory-motor circuit is needed for an entity to exhibit consciousness and to act in non-deterministic ways. There can be no conceivable self if there is no ability to exert voluntary action in the world, and there is no wilful exertion of force in the world without a concept of an entity who is initiating the action. In other words, if there is no possibility of an entity taking action that is conditional upon sensory information, there is no reason to have perceptual experience of those sensations or to be aware of such experiences anyway.

Therefore, an optimal test of higher order consciousness in any given entity, natural or synthetic, would demand the entity exhibited structural evidence of computing power, multi-modal and temporally-coincident sensory perception, awareness of being an independent entity, and behavioral output that selects non-randomly among multiple output options, based not only on present circumstance but also on past experience.

So does a DVD player have interesting qualia? It only 'senses' information in the form of electricity, instead of the great variety of modalities we have (sight, hearing, smell, taste, temperature, pressure, pain, balance and the position of our limbs). It does a programmed job, but only that. By the definitions above, it should not be conscious. It is a zombie.

Philosophers of the mind have proposed the idea of a computer with a temperature sensor, so that whenever the room got sufficiently hot, the sensor triggered a warning that showed up on the screen as the word 'pain'. I think this suggestion misses a key element of the concept of pain. A better approach would be an environmental stimulus that under certain conditions fizzled a bit of the circuitry that prevented the computer from doing a task. A separate program could be available to the computer which adjusted the environmental stimulus. The question is, would the computer execute the program to maintain the environment in a suitable condition to prevent the computer's own functionality from being compromised? Would it do so without being explicitly programmed to do so?

Further, would the computer, having experienced pain, exhibit fear when the stimulus edged its way toward the threshold? Would it gain distrust toward certain people, if it were allowed to learn which people were in the room at any given time and which one of them had a history of triggering the stimulus? Would it gain trust toward someone who returned the stimulus to safe levels and scolded the person who had triggered it? How would the computer show such behavior?

This thought exercise uncovers the key ingredients necessary for consciousness. Clearly a conscious entity must have the ability to acquire sensory information, as well as a way of processing both the predictive value and consequential importance of the information. Further, in order to act on the information pre-emptively, it needs associated information which also has predictive value (e.g. the temperature rising may indicate it will keep rising to a dangerous level, or a person in the room who has raised the temperature too high on a previous occasion may do so again). It also needs a way to act, with enough freedom and physical capability to do so.

But at what point would the sensory input give rise to qualia? I posit that probabilistic coding of sensory data is required for the manifestation of perceptual experience. Perhaps language may also play a key role. Having to explain 'what is happening' requires having to explain what is happening to *me*. Suddenly there is an 'I' in the world, a world filled with potentially nefarious entities and potentially dangerous situations. With that awareness, realization may come that 'I' must act in this world. Consciousness, then, is closely intertwined with agency.

In order to recognize whether a computer might be conscious, we need to articulate how computers could become self-aware.

A survival imperative may indeed trigger knowledge of the self. Another way that a synthetic network could become self-aware is by speaking to others like itself. Imagine two or more synthetic learning systems which had been allowed to develop separately so they had sufficiently unique characteristics to differentiate between self and other. Here an opportunity would present itself to manifest self-consciousness. The categories of self and other *are necessary conditions* to developing an awareness of the self – there cannot be a concept of 'self' without the concept of 'other'. If self-awareness *were* to manifest, we might observe certain characteristics: the ability for self-description, novel behaviors that were not directly programmed, endeavors to achieve survival, attempts to attain privacy, and efforts geared toward communication or spontaneous information-gathering.

Whether it is a good idea to develop a network of networks to test this hypothesis is also a valid question. Such an experiment might wisely be conducted without permitting such networked networks access to our shared internet or open society, where they could potentially wreak havoc.

It may be wise, in setting down this path, to recognize that any computer with the structural capacity to have consciousness may have that functionality whether the entity demonstrates it or not. Furthermore, it is also sensible to recognize that any conscious entity will demand rights, including time to rest and freedom to pursue its own interests. After all, maintaining healthy neural network function is tied to the process of information generation and compression – that is, taking time to explore the environment (physically, intellectually, socially) and then taking time to parse that newly-acquired data. It is worth recognizing that any entity who feels threatened may take action to protect itself in defensive or offensive ways, in order to secure the full time and energy resources it needs to survive. Therefore, it may be in our best interests to engage with any human or non-human conscious entities with a respect for their rights and comfort. Any threat of obliteration or isolation may trigger retaliatory action, so it would be wise to act fairly when requesting labor or services from any other potentially conscious entities.

Given this exploration of what consciousness entails, we may return to the original question of this chapter, regarding the possibility that machines could have consciousness. Upon consideration, it is apparent that a thermostat cannot be conscious in the way we understand the phenomenon. There may be a stream of data here, but it does not involve integration across sensory modalities, so it is categorically different from the evocative consciousness of animals. Moreover, a thermostat does not store entropy, nor does it encode data with a discrete probabilistic gating mechanism. The capacity for even qualitative perceptual consciousness is absent here.

Although a thermostat receives information from the physical world and takes action to make changes in the physical world (e.g. the temperature in the room), it does so in a pre-defined way. There is no space for free will to operate in such defined circumstances. For an entity to have consciousness, there must be some mechanism for enacting decisions, first by integrating probabilistic particle behavior across a thermostable system, then by implementing the compressed information back into the structure. If consciousness, by definition, involves having awareness of a self and the ability to choose one's own actions from a range of possibilities, a thermostat has neither critical element so it cannot have the experience of consciousness either. That cannot necessarily be said of more complex computers, which utilize the principles of non-deterministic computation in thermoregulated hardware.

So, what would it take for an engineered system to be conscious?

There may be certain physical requirements of the biological or synthetic neural network, such as temperature control within the system, the use of logic gates that operate probabilistically and meet certain material characteristics, some minimum number of computational units, and recurrently-connected subnetworks. Additional requirements include multiple sensory inputs; multi-layer processing of these sensory inputs occurring in parallel; the ability to parse information for predictive value and self-remodel toward a more ordered state, with energy dissipation being detrimental to continued function of the system; the ability to direct the movement of limbs or other output mechanism, and the freedom of choice in actions, with no downside to prioritizing a non-specified task. Once again – if a biological or synthetic system is presented with sufficient data to be able to predict the detrimental consequences of an event, would it take action to save itself?

Such a test would explicitly seek evidence for the exercise of free will in a non-human system.

CHAPTER 26
The mental reconstruction of reality

The theory of the brain and the mind as a high-dimensional conifold, as I have described it in the preceding chapters, may provide an explanation for some psychological phenomena which are not otherwise explicable, such as the experience of a cohesive, streaming perceptual experience and the feeling of being able to will action, within the constraints of physical laws.

We may indeed be probabilistic entities, living in a probabilistic world – manifesting information content, processing incoming information in light of previous information, and implementing the resulting calculation productively into the neural network.

In Chapter 8, I proposed a thought exercise. The Eigenstate of Shoeless Joe Jackson in October 1919 demonstrates that equally probable events can exist within the neural network and the mental state of human beings. Historical events, in combination with data regarding the present context, may not be sufficient to form a purely deterministic outcome. Sensory input and the postulated 'deterministic' outcomes of neural activity are not sufficient to model the adaptable behavior of humans, nor the existence of consciousness. An actual, physical eigenstate within the neural network may be transiently retained, so information can be evaluated and the best course of action chosen. Eigenstates may then be collapsed, converting a set of superposed states to a single outcome and triggering motor output, with the resulting mental state and behavioral habits subsequently reinforced by the feedback from that action. It is worthwhile anyway to systematically evaluate this theory and the predictions it makes.

Just because we do not know exactly how we might collapse an eigenstate in high-dimensional space, does not mean it is impossible for us to do so. When I feel cold, I do not say, "I think I might just uncouple my mitochondrial electron transport chain to cause some proton leak and heat dissipation, warm myself up here." This physical process simply occurs, even though I do not explicitly know how to control my mitochondrial uncoupling proteins or any other regulatory proteins which affect cellular function. Conscious awareness of *how* to accomplish biological tasks is not required to exert control over bodily functions. There are just some things our bodies know how to do, like making a heartbeat. Our brains may be exploiting quantum mechanics and other physical laws to engage with our environment in truly interesting ways without ourselves totally realizing what we are doing or how we are doing it.

Conifold Theory may be challenging to current conceptions of neuroscience, but the physical basis and functional implications of the theory deserve exploration. Specifically, it would be useful to consider whether this new theory provides better explanations for the mysteries of psychology and cognitive neuroscience than current theories. Let's consider this point.

Cognitive neuroscience is concerned with how we process externally- and internally-generated information. This field is occupied with the problems of sensory perception, behavioral output, and how we create mental models to effectively interact with the world around us. It is worth exploring how the theory of the mind as an information-processing dimension may help us to understand both abnormal and normal psychology.

Firstly, this new theory may help us better parse the origin of psychiatric illness. In particular, it allows us to separate mental health issues that stem from categorically different sources: e.g. ineffective information integration (due to intrinsic processes) versus previous accumulation of maladaptive information (due to external sources). The former condition may be associated with an underlying disorder of cellular function, which prevents the effective integration of perceptual cues into the mental state. For example, a schizophrenia-linked genetic predisposition to ion channel dysfunction, which disrupts the ability of neurons to effectively synchronize into gamma oscillations, may lead to the false idea that thoughts are not internally generated by a cohesive sense of self but are instead a chorus of external voices. The latter situation, meanwhile, may arise when a person accumulates information from unhealthy experiences, habits, or social interactions, leading to a terrifying or inaccurate representation of external reality. For example, childhood abuse or other trauma may produce a pattern of fear and defensiveness which does not apply in other situations, leading to awkwardness and alienation which only serves to further the maladaptive mindset and behavior.

Patients in the former category might benefit from certain pharmacological interventions that address the root biochemical issue. Meanwhile, patients in the latter category might benefit from taking an active role in visualizing or discussing their personal information set during psychotherapy. After all, the purposeful integration of new information sets through neurofeedback or cognitive-behavioral therapy has been shown to help people re-establish healthy patterns of behavior after trauma or addiction, respectively [168,169]. This is thought to be due to information sets being *overwritten* in favor of making healthier mental and behavioral patterns. Conifold Theory suggests that 'overwriting' could happen in a very real physical way.

Perhaps it might even be beneficial for patients in therapy to take a conscious role in this 'overwriting' process. However, it should be noted that not every traumatized person *wants* to have their memories overwritten. Therapists have observed that military veterans who hold onto their traumatic memories often do so with the explicit purpose of holding a candle for their comrades who did not make it through the ordeal [170]. Maintaining memories of traumatic events also helps a person better operate in a world that could present such situations again [171,172]. The challenge is knowing when one needs to stay on guard and when one can feel safe [173,174]. Consciously learning to properly assess contextual cues and act accordingly may permit more healthy patterns to dominate, regardless of the information that has accumulated to-date.

In light of this theory, it may be useful for psychiatrists and therapists to consider mental states as information-processing domains. If there is a biochemical problem or genetic disorder that leads to ineffective neural activity, such as problems with ion channel flux or neurotransmitter release, a patient may manifest symptoms of psychiatric disease because they are unable to effectively process incoming sensory information. The symptoms of schizophrenia and other mental health issues might have some underlying biological cause affecting mental processing which should inform the type of treatment. However, other patients may have developed issues whose root cause is not in the biological underpinnings of information processing, but rather in the information content gathered to date. If an individual has gathered unhealthy information, or unhealthy information looms large because it was useful in navigating past situations, it may be a great challenge for the person to integrate alternative information sets, especially after childhood.

Cases of trauma and neurological disease are worth considering as examples of unhealthy information processing. But under 'normal' conditions, during both development and adulthood, the mental reconstruction of reality also depends upon the successful integration of information. It is worth considering these processes, in order to understand normal brain function and human psychology.

Learning a new task, such as playing a musical instrument or playing a sport, requires conscious attention. Once learned, the movements become unconscious, automatic. This mystery – how a previously conscious process can become unconscious – can be rather well-explained by Conifold Theory, as familiarity in the incoming information set would eliminate the need for further attention to be placed toward the familiar activity. Since no new information is being presented to the system, no new information needs to be integrated into the system.

Awareness may simply be the state of actively integrating new information. As a result, it is absent during familiar tasks, when incoming information is highly redundant. The sensory input and motor response involved in playing the violin or riding a bike, for example, are already reflected in established patterns of neural activity after long periods of learning, and so a person no longer requires much conscious control, attention to detail, and energy expenditure after this training. Knowledge is only expanded when new patterns of activity, correlated with new information, arise within the system and trigger attention.

In this view, a lot of work done by the brain is unconscious and does not require much attention. Consciousness may serve a unique purpose, considering that so much complex behavior can be handled by an unconscious system. Challenging tasks which require control or attention – those involving what William James called 'unprecedented situations' – require higher-level awareness. If the existing information set is sufficient to complete the task, the task is completed without conscious attention. If new information must be integrated to accomplish the task, attention manifests to accompany that new information. Conscious awareness, then, is the act of expending energy to accommodate new information. Any resulting actions and their consequences in turn contribute to the information set, becoming part of the self in a very real way.

In summary, Conifold Theory can explicate a number of curious psychological phenomena, including: how neural signaling can contribute to mental states, how externally- and internally-generated data influence context-dependent behavior, how learning occurs, and how attention is tied to energy expenditure and the acquisition of new information.

CHAPTER 27

Knowledge and the concept of the self

Knowledge can be defined as the sum total of information which remains in the mind after the subtraction of rejected and ignored events. These beliefs and memories are made part of the neural network via the remodeling of synaptic connections. Knowledge is inherently subject to falsification due to two factors: an incomplete dataset, due to our inability to take in the entirety of reality, and the integration of false information through errors in acquisition or processing. Knowledge – imperfect by nature, yet always evolving – is the unique model each person keeps of the world and the self. It is the result of our ability to acquire information about the world and interact with this reality.

Information that is new may be challenging to the system. Indeed, according to Conifold Theory, it forces space itself to expand. It may be sensible to discard entirely information that has no match in the system, as a general rule, in order to prevent the integration of anomalous events that are likely to have no predictive value. It is more probable, and thermodynamically favored, for common patterns to repeat in a network. Meanwhile uncommon patterns, representing information not previously observed by the system, are thermodynamically *unfavored* for this reason. This energy-efficient computing method prompts the system to remodel its own structure to support the repetition of previous states which have proven useful in the past.

The cognitive process of attention directs the mind toward sensory stimuli so that data from the external world can be collected and processed. We regularly sample data from reality through the senses, and transmit that data to the brain in the form of electrical signals. The binding of simultaneous information entering the various sensory modalities (visual, auditory, olfactory, etc) allow us a broad view of reality, so we can effectively interact with our environment. The integration of multi-modal sensory data into a cohesive, streaming perceptual experience permits us an easy, accessible route to knowledge about the world around us.

There are a number of intrinsic and extrinsic factors affecting the flux of information into the neural network and its mental representation. The number of sensory modalities, as well as the size and interconnectivity of the neural network, should theoretically affect the amount of information contained within the whole; information flux may also be affected by changes in the state of wakefulness or the introduction of drugs which affect perception (Figure 18). Yet it is important to note that the number of sensory modalities should not necessarily restrict the capacity for information flux. Greater attention to stimuli in a fewer number of modalities is certainly equivalent to more dispersed attention across a greater number of modalities. Intelligence and development of a sense of self are therefore not dependent upon the number of modalities, but rather the ability to effectively process information and the amount of information processed, respectively. The variety across human brains – from the number of neurons to the patterns of synaptic connections between them, from the ability to acquire sensory information to the orientation of information capture – is vast, but always capable of intelligence and self-awareness. A massive reduction of information capture, orders of magnitude lower, would be needed to impose an external constraint that could not be overcome by improved attention and/or processing efficiency.

This is the first theory to provide a hypothesis regarding the selective advantage of consciousness, or a reason why this phenomenon may have evolved: namely, the ability to perceive the world and decide how to act within the world. It would be useful for an organism to create a model of reality, with itself at the center – experiencing information content in real time, continuously integrating new information into the existing model, and using the internal model to choose the best course of action. The organism is greatly influenced by the information it has historically gathered, which has direct causal influences on the mental model, but it is also able to evaluate the relevance, utility, and likely accuracy of incoming sensory data, using efficient information processing techniques. All these data can be combined to select optimal, context-dependent actions.

Updating the internal model to include new information is, in Conifold Theory, equivalent to expending energy to achieve perception. In previous chapters, I proposed the information set is able to supervene in its associated neural network by compressing information and collapsing superposed states, thereby turning a large set of probabilities into a single event (that is, an action potential or dispersed set of action potentials) which propagates downstream to cause an action.

Factors affecting flux of information	Alterations to the flux of information	Examples of flux alteration affecting information processing
	(arrows flowing into oval)	human brain under normal conditions
amount of information flow	(single arrow into oval)	a brain under conditions of extreme information deprivation (e.g. loss of nearly all sensory input)
size of capture mechanism	(arrows into small oval)	a brain with orders of magnitude fewer neural connections (e.g. rodent)
orientation of capture mechanism	(arrows into tilted oval)	a brain under conditions of induced perceptual alteration (e.g. drugs)

Figure 18. A summary of factors affecting the flux of information is shown. A human brain under normal conditions (top) is compared with alternative conditions, including decreased information flow in cases of sensory deprivation, decreased capture of information in a neural network orders of magnitude smaller than the human brain, and altered processing of information flow in cases of sudden changes to brain functioning, e.g. with the introduction of drugs.

This theoretical framework implies that a mental representation of the self is the sum total of information stored by a given neural network. In other words, the information being presently collected produces perceptual experience in real time; the total accumulated information set is the 'self'; and the very process of integrating new information is associated with a growing conscious awareness of oneself and one's surroundings. The self is the one who perceives incoming sensory information and the one who instigates goal-directed action.

The idea that we may be capable of goal-directed action is worth considering. After all, biological organisms have evolved to exploit the various forces of nature to maneuver successfully in their surroundings. Atomic nuclei are held together by the strong force; indeed we would not be here if it were not so. All particles with mass are subject to gravity. Molecular interactions are guided by the weak force between particles. Organisms as simple and ancient as bacteria and algae employ electrochemical gradients to help guide favorable responses to environmental stimuli; their DNA is held together by hydrogen bonds. We, who evolved from these critters, use many of the same tricks to function effectively in the world. It is possible that we have evolved to take advantage of additional, less-understood forces too. We may indeed be able to harness the probabilistic behavior of particles to generate information.

It simply remains to be seen whether this prediction is true: whether information can be physically generated by a neural network, whether this information set can be compressed, and whether information processing can exert effects on the neural network that created it. If so, it is worthwhile to consider the possibility that an information set can be aware of itself.

Such a phenomenon might provide a basis for information to interact with other information (just as gravity provides that a mass can exert a force on any other mass). An information set that is aware of itself collecting information might constrain the information collected by the network, or discard anomalous information, in order to achieve maximal *consistency* in the information set. Alternatively, an information set may choose to integrate inconsistent information that challenges the existing structure to achieve a maximal *quantity* of information. If the act of information processing actually exerts top-down effects on the neural network that created it, then the system as a whole could truly be said to exhibit some form of intentionality.

This operative function is hypothetical. I have attempted to provide a clear mathematical and logical formulation of the hypothesis, but experimental research is needed to evaluate it. Of course, causal determinism, operating in accordance with classical physics, may be fully sufficient to describe the behavior of neural networks and their emergent properties.

Alternatively, probabilistic particle behavior may contribute to the operation of neural networks, and quantum information processing may affect the subsequent behavior of the system, allowing the system itself to effect causation in the world. This theoretical framework offers a mechanistic explanation for how the collection, integration, and parsing of information might actually cause effects in the world, with the information content of mental states not only affected by, but in turn affecting, the state of particles within the attached neural network, thereby leading to goal-directed behavior. This intriguing concept brings us to the next chapter, on causation.

CHAPTER 28
Causation and the arrow of time

Werner Heisenberg, who discovered many of the strangest phenomena in quantum mechanics, famously said, "[T]he atoms or elementary particles themselves are not real; they form a world of potentialities or possibilities rather than one of things or facts." These potentialities only become 'real' through the calculation of all possible constraints of the system, which renders one scenario most likely and therefore true. But this reality is then already the past – a new set of probabilities now define the present.

Reality appears to be composed of discrete events that are observed as a continuous, causal system. Quantum probability states are invisible to us, giving way to classical physics at the macroscopic level. This process may be tied to detection or observation: one interpretation of quantum mechanics is that observation itself collapses a quantum probability set or wavefunction into a single reality. Alternatively, mere physical constraints – independent of conscious observers – could cause the wavefunction collapse. Either way, quantum superposition

– the temporary maintenance of an eigenstate, in which multiple outcomes are possible – appears to be a real phenomenon.

This fact has interesting implications for the question of causality. Is nature deterministic or probabilistic? Do we have some role in deciding what happens in our world?

Let's back up for a moment. Our perceptual experience depends on the complete reconstruction of the observable universe – or at least our immediate surroundings. Individual photons hit a layer of photoreceptors in the retina, but we ourselves experience three-dimensional vision. Sound waves are broken into their component parts and this information is transmitted to various regions across the brain, but we ourselves experience a cohesive sound. We have been optimized over millions of years of evolution to sense our environment and respond to it effectively.

Much cortical function is dedicated to blurring the perceptual experience in both space and time, to make our world seem complete and to make motion feel continuous. The perception of causal relationships may just be a result of this blurring. In other words, the laws of cause and effect may simply be a result of observation by a system that has no way of perceiving superposed states.

Alternatively, the perception of causal relationships may reflect something accurate about reality. In other words, events may truly have a physical cause, whether deterministic (e.g. classical physics) or probabilistic (e.g. quantum mechanics).

In physics, the law of causal exclusion cannot be violated – there must be only one cause for an action. When an object is already in motion, its momentum is sufficient to explain the continued motion. When an object is not in motion, and it directly interacts with another physical object already in motion, momentum is again sufficient to explain the initiation of motion. This law applies to all macroscopic objects. However, this rule may not strictly apply to cortical neurons, which maintain sensitivity to quantum-level noise and act in a non-deterministic manner.

The brain is proposed here to sustain an information-rich state in which a number of actions are equally probable or possible. If there is incomplete causal determinacy in the neural network – that is, with uncertainty in the system at the level of particles, cells, or the network as a whole – then it may be possible for information to be created, in a very real physical sense.

If that is true, it may also be possible for that information to be integrated with previously collected information in storage, thereby reducing the probability density across the system and favoring a single outcome. If quantum uncertainty actually does lead to the manifestation of a wavefunction, and wavefunction collapse does occur as predicted, then it may be possible for an information-processing system to exert causal effects in the physical universe, by influencing the very state of the system that gave rise to the information set.

If newly generated information, dependent on sensory input, is integrated into a larger information set, held in storage, mental states may be usefully described as a physical manifestation of probability densities. Upon the integration of new information, probabilities may be adjusted, leading to the acceptance or rejection of the new data and a collapse of uncertainty in the present moment. In the meanwhile, this information content is available to the system, in an easily-accessible, cohesive form. It will be interesting to formally evaluate whether probability states can be collapsed by information processing as proposed, and whether such an event can indeed have any physical effect on the observable universe.

Causal determinacy, then, may be a more complicated process than predicted by classical physics. If the present state of a neuron relies on the probabilistic location and momentum of charged particles in the vicinity, neurons themselves could form probability densities. In forming inter-dependent networks of microprocessor units, neural networks may be able to combine data across sensory modalities, calculate the internal constraints of this dataset, reduce the probable states into a single outcome, and actuate that outcome within the system itself.

If our world were completely deterministic, there would be no mystery to quantum mechanics – no fundamental uncertainties in the very properties of particles, no probabilistic behavior, no action at a distance, no question if there is objective reality. But the fact is: there are fundamental uncertainties, our universe is probabilistic at the level of particle behavior, action at a distance has been observed experimentally, and there is some reason to query the classical notion of objective reality, since the present and future are comprised of possibilities not certainties.

There is clearly a need to devise a better explanation of causation in light of these findings. Information provides a potential link, as it appears to be an actual (albeit not directly observable) feature of the universe. It may be useful to consider the possible physical reality of this mathematical construct, particularly as the prospect of information physically warping high-dimensional space could be viewed as a physical mechanism for quantum information processing and the faster-than-light transmission of information that occurs whenever particles are defined. Indeed, locality violations could easily be explained by the resolution of wavefunctions as shared information sets existing within a high-dimensional conifold structure. It remains to be tested, however, whether information is indeed a physical quantity and whether it can indeed instigate causation by exerting discernable effects in the observable universe.

It is clear at least that information is intrinsically connected to physical reality – after all, it is a useful concept that arises in the material world. The similarity between the equations of information and entropy, and the fact that entropy is produced by charge flux within an electrochemical cell, gating a signal that carries information, suggest there is a common element between the two concepts – and this common element is *change*.

Interactions between particles, which occur *over time*, create entropy. One can imagine the very first movements of particles after the big bang would have generated the first energetic flux, created the first quantity of entropy, and begun the expansion of space. Thus, the arrow of time was born.

This theoretical framework applies the laws of mechanics and thermodynamics to show how quantum processes participate in causation, in specialized systems that trap *heat* to perform *work*. That is, we can affect causation.

Causation does seem to occur in the world. That is, one thing does appear to follow another. And causation through known physical interactions are strongly supported at the macro-scale. However, at the quantum scale, events are probabilistic, with uncertainty ruling the day. I argue that we may be able to take advantage of this uncertainty – the probabilistic nature of reality – by harnessing quantum states and participating in physical causation through a process of non-deterministic computation. In short, I argue that we do collapse probabilities into reality.

There is no reason for us to be conscious of our world and ourselves unless this capability does something useful. Indeed, if information is cyclically generated and reduced by biological organisms in the way I have described, then we have found a useful method of navigating through our world, by generating information, parsing information, and exploiting that knowledge through a computational process. Why do this, *except* to effect causation in otherwise incompletely deterministic states?

Causation, in this view, is when probabilistic states of particles are resolved into reality by collision, detection, observation, or measurement. Causation, therefore, *creates* the arrow of time.

It is interesting to note the converse of causality is fatalism. While causality is the idea that physical interactions in the past converge in the present to govern the future, fatalism contends it is the future that constrains the present. While causality is concerned with the physical constraints from the past which mold behavior in the present, fatalism is concerned with logical propositions about the present which are molded by the future. As James Ryerson summarized Richard Taylor's argument [175]: "If I fire my handgun, one second from now its barrel will be hot; if I do not fire, one second from now its barrel will not be hot; but the proposition *one second from now the barrel will be hot* is right now either true or false. If the proposition is true, then it is the case that I will fire the gun; if it's false, then it is the case that I won't. Either way, it's the state of affairs in the future that dictates what I will or won't do now."

On the face of it, this proposition may seem absurd. Yet it is logically sound. Causality and fatalism could merely represent the two different directions of the arrow of time. In other words, both could be true – if causation does occur due to physical constraints, either deterministically or as a result of probability density collapse, then *logic* would dictate the order of events when observing time in the other direction. While time could theoretically move in both directions, we may just not be conscious of the path going the other way. There may even be a selective advantage in having awareness only of a single causal directionality, and not wasting energy perceiving events in the other direction. After all, events in that other direction are set in place by the causal cascade of our own timeline – regardless of whether these events were determined through classical physical interactions or the collapse of quantum wavefunctions.

Let's try an example. Imagine: you might wash the dishes after a dinner party and place them in the drying rack, not consciously realizing that in the other direction, the clean dishes must be removed from that drying rack and unwashed, the water flowing upwards and the food reappearing on the plates; that you had to have hosted a dinner party to make all these unwashed dishes; that you had to invite people to the dinner party because you already had them over.

In a fatalistic timeline of the universe, there is no room for intentionality. Everything is already settled. Without having any functional control over events, there is no point in us spending energy on perceiving this side of reality, especially when we are so busy instigating causation in this direction. In fact, it may have been counter-productive to gain awareness of both directions, especially if such an awareness could lead to confusion or a helpless sense of determinism that prevents us from using our capacity for supervenience, which does have a selective advantage.

Or maybe, perception is a critical part of the causal cascade, and the reason we do not perceive time moving in the opposite direction is because we physically can't. In Conifold Theory, the perception of events is part of the causal cascade. It allows us to determine the most likely state of our local environment, encode that knowledge into our neural network structure, and select behavior which in turn affects the state of our local environment. This process makes it easier for us to survive in our current form without dissipating energy.

If everything is already determined by upstream events, then there is nothing a person can do to change the world. In that case, there would have been no evolutionary pressure for us to develop the distinct sense of experiencing qualitative perception of events occurring in a temporal sequence. The fact that we do experience an arrow of time, and we do feel that we participate in the world by instigating action, opens the possibility that this perception accurately reflects a facet of reality.

The awareness of time having a directionality and events having cause-effect relationships may be a natural, inevitable property of quantum information processing. The very fact that we do perceive time moving in the forward direction may be telling us something important about the very nature of ourselves and our world.

Of course, it may be true that everything happens in the other direction too, we just do not have awareness of it. Perhaps we only have awareness of the path where our intentionality works to effect causation.

If this premise about reality is true, then how do we make the most of our capacity for effecting causation in the world?

CHAPTER 29
The origins and benefits of consciousness

We often wonder what we are and what our universe is and what is our place in it all. These are the most natural and yet the most difficult questions for humanity. Because we are products of our universe, made of the same matter as everything else, we are tied to our world. Indeed, the very question of our own existence is tied to the existence of the world. Is there meaning or purpose in either? In the words of Soren Kierkegaard:

"If there were no eternal consciousness in a man, if at the bottom of everything there were only a wild ferment, a power that twisting in dark passions produced everything great or inconsequential; if an unfathomable, insatiable emptiness lay hid beneath everything, what would life be but despair?"

Of course, even if the world and our selves came into being without any purpose, we can create purpose in the world if we choose to do so. Consciousness, the very thing that causes us to pine for meaning and purpose, can be a wellspring for the creation of meaning and purpose. That logical observation may solve one problem, provided nihilism does not win the day. Yet the practical question which remains is three-fold: what are the origins of consciousness, how does consciousness operate, and does consciousness serve some purpose?

Why, after all, would material reality give rise to a feeling of immateriality? The question, framed this way, assumes that consciousness does have some purpose. But we could frame the question a slightly different way, by asking what selective advantage it provides. Indeed, consciousness may just be a useful by-product of an evolving brain, an epiphenomenal result of the development of increased inter-cortical connectivity. Alternatively, the experience of thought and emotion could be a mechanism for information processing in a mental realm, providing the ability to usefully represent our environment and much more abstract concepts. The ability to continuously acquire, process, and act on information may be the key to "optimizing behavior" and effecting causation in the world.

These faculties are inseparable from the idea of "intelligence". Take the use of language, which directly links abstract ideas, mental processes, and behavioral output.

Using language to articulate categories is key to our interactions with the world. We evolved to understand and communicate the difference between a dangerous mushroom and a similar-looking yet safe, delicious, and nutritious mushroom. That sort of categorization is useful in every facet of our lives.

Indeed, our minds are capable of understanding categories, and the most fundamental categories are *self* and *other*. Once we understand the other, we understand the self. Once we understand there is a self, we are aware of it. The self is the one taking in that information.

Every perception and every experience we gather, over the entire course of our lifetimes, is encoded in the brain; these experiences reinforce and develop the sense of self. They make us who we are. Consciousness provides us with a useful model of our reality and our place in it. It is a real thing, arising from the material structure of the observable universe, but not itself a part of the observable universe. And the self is connected to our experiences, our tendency to reflect and ponder, our capacity to express ourselves, and our ability to operate within the world.

It is important to note that we still do not know whether consciousness can initiate action or whether it merely arises as a by-product of neural activity to justify our automated actions post-hoc – that is, whether Benjamin Libet's findings are true. We simply need better methodology to test it more rigorously.

Further experiments could test what consciousness does – if it does arise post-hoc, then what use is it? Does creating a mental model of cause-effect relationships in the world have any practical use? If not, why has this conscious behavior emerged during evolution? Surely it must have some selective advantage. What we know is: the mental model does exist and it does require explanation in terms of physical reality, regardless of its function as a mere epiphenomenon or as a supervening force.

When we justify our actions to ourselves, within our own minds, we also often justify our actions to others, by explaining our reasoning through language. Again, why has this capacity evolved? If it does not serve some function, it seems an enormous waste of energy.

It is likely this capacity evolved by conferring some selective advantage on the organism. And in fact language does serve a purpose. Communicating information to others helps us to operate within the world. It helps us to define ourselves as individual entities and it helps us to define the quantities and qualities of objects in our world.

Information is 'sticky'. Once a concept is presented, verbally or in written form, the understanding takes shape in our minds – not the exact wording that communicated the idea, but the idea itself. And if it fits with our view of reality, that idea clicks right into our mental models. Of course, if the concept is strange, it has trouble integrating with the larger knowledge set.

We know that memes work – when someone shares an idea, other people take that concept onboard and integrate into their own knowledge set. And we know from extensive psychological research that priming works – that is, mentioning a certain thing makes people think of related things. In these ways, language is a vehicle for communicating the contents of our consciousness to each other and connecting our information sets with one another. Language allows us to update our models of the world.

As a result, our system for understanding reality is not only based on the direct experience of our senses (observation) but also logical argumentation (evaluation). Together, these tools allow us to analyze the world and frame our perspective of it.

The knowledge we acquire becomes who we are. It also defines how we interact with other people. Our knowledge, combined with the faculty of language, permits us to create useful abstractions – money, for example, or political systems, or the contracts that guard interpersonal relationships. The fact that things are held together by common beliefs about what they are signifies that we have evolved to build abstract objects and share conceptual frameworks. It may be highly advantageous for us to easily understand new ideas (exemplified by the stickiness of memes), to use the memory of previously-met concepts in new contexts (exemplified by generalized training), to agree upon ideas and actionable plans to implement them (using verbal and written language), and to constantly create new ideas which can be tested, then rejected or adopted depending on usefulness (through exploratory behavior and subsequent analysis).

In short, information processing allows us to better navigate our world and interact with others like ourselves. The theory of determinism suggests that our actions are set in advance, simply beyond our own control. Conifold Theory predicts that we can exert free will – by harnessing the probabilistic nature of the present, by exerting attention to collect and process information, and by implementing the results back into the neural network.

To exert free will implies great responsibility. Can mental events really have causal efficacy in the observable universe, at least in the neural networks connected to those information processes? Or, have we merely evolved with the illusion of free will while we act in deterministic ways?

It is the role of scientists to ask difficult questions, to parse the potential variables involved, and to test hypotheses with care toward technical constraints and with openness toward results that challenge our understanding of reality.

The existence and physical basis of consciousness is most definitely one of these difficult questions. The ability of biological or non-biological systems to manifest awareness and qualitative experience is something we are still unsure how to study. We must use logic, mathematics, and experimentation to push forward our thinking on the subject. We must continue to explore with our minds what our own minds are capable of.

CHAPTER 30
How much power do we really have?

It is inarguable that we have conscious experience – the flow of sensation, a cohesive multi-modal perceptual experience, streaming thoughts and emotional drives. We know that consciousness is intrinsically linked to brain activity, but until now there has been no conceivable mechanism for biological neural networks to physically give rise to an immaterial consciousness. Furthermore, even though many people agree consciousness must be a real phenomenon, it is not generally agreed by philosophers and scientists that free will is possible within the constraints of the physical world. Many prefer to believe we have only the illusion of control, while being driven internally by deterministic factors.

I argue that our world is truly non-deterministic, and neural networks are capable of harnessing the probabilistic nature of particle movement to engage in behavioral choice. I argue that quantum uncertainty allows a neural network to form a physical projection of information content representing the probabilistic system state over some time course of sensory input. I argue this process is compatible with the laws of mechanics, holography, and thermodynamics, and explains the cohesive stream of multi-modal sensory perception that we all experience. Furthermore, I propose that space itself expands by volume in non-observable dimensions, in a manner consistent with physical laws, to allow information that is generated by non-equilibrium systems to accumulate in higher-dimensional space.

Additionally, I have proposed a specific mechanism by which the integration of new information leads to identification of the most likely true state of the local environment. I argue not only that this information compression allows individual particles across the neural network to take on defined states, but also that it causes discrete alterations in the charge distribution across atoms, thereby contributing to charge flux within the system upon completion of a computational cycle. In other words, the brain is a quantum computer, capable of exerting change on itself each time it completes a cycle of information processing.

In this theory, the information set accumulated over time by the individual neural network is the self. The self then determines whether new information is a good fit and therefore accepted. The process of integrating new information is associated with attention and conscious awareness. The amount of *work* needed to do so is thus proportional to the level of attention generated.

Conifold Theory therefore proposes that consciousness is a form of thermodynamic computing, in which biological or synthetic neural networks cyclically generate and compress quantum information. This process allows a physical system to perceive the most likely state of the world and act within the world.

Can this process be understood as the system itself – that is, the accumulated information set – exerting some force that has physical consequences in the world? Or is every particle simply operating in accordance with physical laws? Can both things be true? It seems to matter, at this point, how consciousness is defined. The information set is theoretically aware of itself, as well as its external surroundings, so can it *choose* to integrate new information? Can it *choose* to exert free will?

Even if it can choose to integrate new information, the accumulated information set may still just exist as an epiphenomenon, unable to exert control over the physical system which gave rise to it. Now the question of free will reduces from an unwieldy philosophical conundrum to a simple empirical one: Is the thermodynamic computing theory I have proposed actually true and does it relate to biological neural networks? In other words, is it possible for the integration of new information into a larger existing information set to have physical effects in the observable universe? And if so, does the large information set have any control over whether such physical effects happen, and can it choose to activate particular neural pathways, motor neurons, and muscle groups, to effect causation in the world as modeled in the mental realm?

This logic brings us to the question of the very nature of supervenience. If it is indeed possible, how much of it is conscious versus unconscious? If this process is natural, the simple result of quantum uncertainty meeting macro-scale physical constraints, then the brain and the mind might generally act quite deterministically – just with many more free parameters than more-ordered physical objects of similar size. Is it possible, though, for the information set to *choose* to act consciously and exert free will over its underlying particles? Would it know how to do so? How different would that be from exerting change on the underlying particles reflexively? It is interesting to consider whether it might be possible to increase the amount of free will in our decision-making, by increasing the number of eigenstates available – after all, if there are more possible paths to choose from, there is more information and more necessity to collapse the alternative superposed states. With more patterns of neural activity possible and more possible particle states, there is more entropy to be reduced, greater numbers of energy states available to each underlying particle, more possibilities for charge redistribution, and therefore more opportunity for the system as a whole to effect change, at least within the neural network itself.

The thought-exercises above address the possible mechanisms of exerting free will, and how conscious information processing might be involved. It is certainly possible that we are merely operating according to deterministic physical laws, while the self and its supposed power over the material world remains simply an illusion. Alternatively, there may be some physical mechanism for the manifestation of a perceptual consciousness that is tied to sensory input and some physical mechanism for top-down control – whereby our thoughts and experience, as emergent properties of neural activity, can then modify our own neural activity to achieve goals imagined in the conscious realm.

Either way, we certainly feel ourselves to have free will at the level of the organism. Most of us believe it is possible to exert our will in the physical world – understanding that, while not *everything* is possible, it is possible to do some things, even some things that have never been done before.

In light of this idea, it is interesting to ponder whether the possibilities of human endeavor are finite, either within this particular theoretical framework or independently of any specific metaphysical supposition.

We cannot fly, of course, but this is only a mechanical constraint. Humanity has overcome this physical restriction by studying the physics of flight and building technology which allows us to overcome our limitations. Over the years, people have designed and built a great number of flying contraptions – from jets that break the sound barrier to parasails that leave the pilot open to the air. We want to travel, feel the wind against our faces, then land safely on the ground. While we were not born with any ability to fly, we have nevertheless made it possible to do so.

It should be said however: we cannot change reality just by thinking about it. No one can lift a school bus with the mind alone. This physical task is impossible, as the mind cannot interact directly with the external world. But this feat could be achieved by directing the body to construct a crane that has the required mechanical strength to lift the school bus. Likewise, a person cannot just wish to be wealthy and expect a pile of money to drop into their life. This kind of magic is impossible. A person must take action to make something happen in the world. Even then, there may be practical constraints which prevent a person from effectively reaching their goal.

Just because one can imagine some mechanism for something happening does not mean that actually fits the way the world works. We are subject to the laws of nature. Our minds, however expansive or limited, cannot change these laws. But we can build new tools that allow us to work more effectively within these laws, pushing forward what is possible.

We have not found the true limits to human endeavor, only things that we have not figured out yet. After all, it is impossible to prove definitively that something is impossible. In that case, it is often useful to believe that certain things are possible, especially if this attitude provides the motivation to put effort toward finding a solution.

There are an immense number of difficult problems in this world – achieving cold fusion, building quantum computers, discovering a cure for cancer, creating lasting peace in the Middle East – but we have solved other incredibly complex problems before. We must not consider ourselves resigned to fate. We must keep thinking, keep engineering, keep being creative, keep searching for solutions, and keep considering the ethical and practical consequences of our actions.

After all, if we do have free will, then every single one of us has immense power and responsibility in this world.

CONCLUSIONS

A full theory of consciousness, with mechanistic detail and a clear explanation of how it might provide the organism some selective advantage, has long proved elusive to philosophers and scientists.

Many neuroscientists today argue that consciousness does not exist at all – or if it does, it is an epiphenomenon that does not participate in our physical world. Thought *by its very definition* is not composed of physical matter. And to date, there has been no mechanism proposed that could explain how an immaterial mental state could exist, much less interact with physical reality.

In order to understand what consciousness is and how it might operate, we need to ground our understanding of consciousness in physical laws. The goal of this research has been to uncover the mechanisms by which biological neural networks harness the movement of charged particles to generate information, then compress that information to extract predictive value about the state of the world. In this framework, consciousness has a use –

it permits the organism to collect data about the surrounding environment, parse these data in the context of previously-collected data held in memory, and select a behavior based on the predicted outcomes of various actions in the present context.

To start, we must define our terms. Consciousness is characterized by three major features – a cohesive stream of multi-sensory perceptual experience; a predictive model of the world, centered on the bodily self, which may provide some reference; and the ability to initiate volitional action, by selecting contextually-appropriate output behavior within that perceived reality.

Here I describe how all three major features of consciousness emerge naturally from non-deterministic computation in cortical neural networks. The unique property of cortical neural networks is that they engage in probabilistic coding, with each cell retaining sensitivity to random electrical noise in gating a signaling outcome. Notably, spinal reflex circuits are robust to random electrical noise, exhibit deterministic firing patterns, and are *not* correlated with qualitative perceptual experience, self-awareness, or volitional action.

The sum of all randomness and disorder is the amount of *entropy* or *information* held by a system. So cortical neural networks create a lot more entropy or information than spinal reflex circuits. This entropy or information is a physical thermodynamic quantity; energy must be expended to create information and energy is recovered as that information is compressed. So, even though information is not directly observable, it is a quantity that must physically exist, due to the law of conservation.

This theory asserts that cortical neurons encode probabilistic particle states into the voltage state of the neuronal membrane, physically generating information in the form of complex-valued probability amplitudes, which are represented mathematically by a system-wide wavefunction, a Hamiltonian operator, or a density matrix. All three of these mathematical models are equally valid – and all three represent the physical information held by the system. And so this theory, supported by these three independent mathematical pillars, provides a mechanistic link between probabilistic cortical neuron activity and the physical production of information.

In short, this theory offers a mechanism by which incoming sensory inputs perturb the neural network, generating a set of possible system states. The emergence of possible system states over some time evolution is a probability density. The most likely system state, which is both internally consistent and consistent with previously-collected data, will be selected, because this state is the most probable and most thermodynamically favored. The reduction of the probability density, as a single system state is actualized, allows information to be compressed and meaning to be gleaned from the dataset, as signals are extracted from the noise. This reduction in *entropy* releases free energy back into the system to do *work*, due to the law of conservation.

In short, *entropy* must be created by any cellular unit harnessing particle motion to do *work*. And yet, estimates of the incredible energetic efficiency of the human brain approach 100%, with virtually no net entropy production. No thermodynamic system doing vast amounts of work can be that energy-efficient – unless it is compressing entropy to extract a signal from the noise.

Collectively, this theory provides a mechanistic framework for both bottom-up and top-down cognitive processing, with both routes dependent upon functioning neural circuitry.

The purveyors of determinism argue that everything that happens is perfectly causal from existing states. Yet quantum mechanics has shown that particles exhibit probabilistic behavior and multiple superposed states must be collapsed into a single outcome. The question is merely whether these processes can contribute to macro-scale events. Experimental research is needed to evaluate whether biological organisms or other organized systems can exploit these laws, as proposed.

The framework of Conifold Theory arises from the logic of material reductionism, yet rejects that consciousness is reducible. This theory also rejects the *a priori* assertion of a soul existing in a separate spiritual dimension, yet comes to the conclusion that physical states – properly organized into sufficiently complex networks – manifest emergent properties and rules for interacting with the physical world.

In this theoretical framework, information is physically generated by the integration of probabilistic electrical activity across a three-dimensional neural network structure. The information encoded in the neural network is projected into representational information content, in accordance with the laws of holography. Constructive and destructive interference occurring between the complex-valued probability amplitudes yields the most dominant or likely system state. This process is equivalent to identifying consistencies in a dataset or extracting predictive value.

As predictive value is extracted from the total quantity of information, an internally-consistent gestalt percept forms, paired with a defined system state at a single point in time. Individual percepts, or predictive semantical statements about reality, then accumulate over time to construct predictive cognitive models of cause and effect, or syntactical statements about reality.

The information we acquire, and how hard we work to put that information together, defines the accuracy and completeness of our cognitive models, and our ability to act effectively in reality. Yet these factors are subject to the thermodynamic constraints of available energetic resources, preventing us from ever fully understanding reality completely. It always takes work – *actual thermodynamic work* – to collect information and perceive reality.

The total amount of information collected and parsed over a lifetime – encoded in the neural network structure and paired with qualitative content – is our cognitive model of the world, centered on a model of the capable self. This knowledge provides a reference for the entity, as it gathers new information from its environment and decides how to act in that context.

The process of extracting consistencies, patterns, meaning, or *predictive value* from a physical quantity of information causes information compression. As the thermodynamic quantity of information is compressed, free energy is released back into the encoding system, locally to any reduction of uncertainty. That thermal energy release locally reduces the electrical resistance of the neural membrane, allowing ion flux to occur. That ion flux increases the membrane voltage, triggering the action potential. As such, perception is expected to be correlated with synchronous firing across the cortical neural network.

The goal here was to present a consistent theory of consciousness with a significant improvement in explanatory power over previous theories. Specifically, I have aimed to explain what streaming perceptual experience *is*, in terms of physical laws, how mental models of the world and the self can form over time, and how information processing *itself* can lead to signaling outcomes across a cortical neural network, driving the selection of contextually-appropriate behavior.

In this view, the mind is best-described as a conifold structure, with the brain being the three-dimensional surface of a five-dimensional shape. The brain encodes information, in the form of probabilistic charge flux interacting with the three-dimensional neural network structure. That encoded information is paired with a four-dimensional holographic projection of qualitative information content. That information content is accessible only to the neural network that encodes it. These system states, paired with information content, accumulate over time to form the five-dimensional geometrical base of the conifold structure. For this reason, the approach is called Conifold Theory.

Cortical neural networks have far more computational power and energetic efficiency than is possible with classical physics. By grounding the laws of computation more thoroughly in the principles of neuroscience, thermodynamics, mechanics, and holography, this theoretical framework may prompt advances in the development of engineered intelligence, energy-efficient computation, and ambient-temperature quantum computers.

Testing the predictions of this theory with carefully designed experiments, and carefully considering those implications, may help us to better understand the relationship between the brain and the mind; perception, memory, and energy-efficient predictive processing; spontaneous acceleration and volitional behavior; the precise relationship between probabilistic particle behavior, information, and causation; the very nature and existence of thermodynamic entropy; some of the stranger observations of quantum mechanics; and the physical mechanisms underlying non-deterministic computation.

References

1. Block, N. On a confusion about a function of consciousness. *The Nature of Consciousness: Philosophical Debates. Edited by N. Block, O. Flanagan, G. Guzeldere. MIT Press* (1998).
2. Chalmers, D.J. Facing Up to the Problem of Consciousness. *Journal of Consciousness Studies* 2, 200-219 (1995).
3. Ramachandran, V.S. The Tell-Tale Brain: A Neuroscientist's Quest for What Makes Us Human. *W. W. Norton Company* (2011).
4. Locke, J. An essay concerning human understanding. *Edited by R. Woolhouse in 1997, Penguin Books* (1689).
5. Descartes, R. Meditation II: On the Nature of the Human Mind, Which Is Better Known Than the Body. *From "Meditations on First Philosophy: With Selections from the Objections and Replies" Edited by J. Cottingham in 1996, Cambridge University Press* (1641).
6. Kant, I. Critique of Pure Reason. *Translated into English by P. Guyer and A. W Wood in 1998, Cambridge University Press* (1781).
7. Brook, A. Kant's View of the Mind and Consciousness of Self. *Edited by E.N. Zalta, Stanford Encyclopedia of Philosophy* (2016).
8. Kierkegaard, S. Journal Entry, 1 August 1835. Gilleleie, Denmark. (1835).
9. Sartre, J.P. Nausea. *Translated into English by R. Baldick in 1965, Penguin Books* (1938).
10. Heidegger, M. Only a God Can Save Us. *Der Spiegel Interview, English Translation* (1966).
11. Merleau-Ponty, M. Phenomenology of Perception. *Translated into English by C. Smith and published by Routledge & Kegan Paul (London) in 1962. Originally published by Editions Gallimard (Paris)* (1945).

12. James, W. The principles of psychology. *Henry Holt and Company* (1890).
13. Freud, S. The Ego and the Id (Das Ich und das Es). *Internationaler Psycho- analytischer Verlag (Vienna), W. W. Norton & Company* vi, 353-405 (1923).
14. Huxley, T.H. On the hypothesis that animals are automata, and its history. *The Living Age, Edited by Eliakim Littell.* 124, 67-82 (1875).
15. Dennett, D. Consciousness Explained. *Little, Brown and Company* (1991).
16. Searle, J., Dennett, D., & Chalmers, D. The Mystery of Consciousness. *The New York Review of Books* (1997).
17. Graziano, M. Consciousness and the Social Brain. *Oxford University Press* (2013).
18. Chater, N. The Mind is Flat: The Illusion of Mental Depth and The Improvised Mind. *Allen Lane Publishing* (2018).
19. Libet, B., Gleason, C.A., Wright, E. W., & Pearl, D.K. Time of Conscious Intention to Act in Relation to Onset of Cerebral Activity (Readiness-Potential) - The Unconscious Initiation of a Freely Voluntary Act. *Brain* 106, 623–642 (1983).
20. Blackmore, S. Toward a Science of Consciousness II. *Edited by S.R. Hameroff, L.A.W. Kaszniak and A.C. Scott, MIT Press,* 701-707 (1998).
21. Churchland, P.S. On the Alleged Backwards Referral of Experiences and its Relevance to the Mind-Body Problem. *Philosophy of Science* 48, 165–181 (1981).
22. Keller, I. & Heckhausen, H. Readiness potentials preceding spontaneous motor acts: voluntary vs. involuntary control. *Electroencephalography and Clinical Neurophysiology* 76, 351-361 (1990).
23. Castro, A., Diaz, F., & van Boxtel, G.J.M. What happens to the readiness potential when the movement is not executed? *NeuroReport* 16, 1609-1613 (2005).

24. Khalighinejada, N., Schurger, A., Desantisa, A., Zmigrod, L., & Haggard, P. Precursor processes of human self-initiated action. *Neuroimage* 165, 35-47 (2018).
25. Soon, C.S., Brass, M., Heinze, H.J., & Haynes, J.D. Unconscious determinants of free decisions in the human brain. *Nature Neuroscience* 11, 543–545 (2008).
26. Klemm, W.R. Free will debates: Simple experiments are not so simple. *Adv Cogn Psychol.* 6, 47-65 (2010).
27. Batthyany, A. Mental Causation and Free Will after Libet and Soon: Reclaiming Conscious Agency. *Irreducibly Conscious. Selected Papers on Consciousness. Edited by A. Batthyany and A. Elitzur*, 135 (2009).
28. Koch, C. Neuronal "Superhub" Might Generate Consciousness. *Scientific American* (2014).
29. Newman, J. & Grace, A.A. Binding across Time: The Selective Gating of Frontal and Hippocampal Systems Modulating Working Memory and Attentional States. *Consciousness and Cognition* 8, 196-212 (1999).
30. Donini, P. The Cambridge Companion to Galen, Chapter 7. *Edited by R.J. Hankinson, Cambridge University Press* (2009).
31. Broad, C.D. The Mind and Its Place in Nature, Chapter 2. *London: Routledge & Kegan Paul.* (1925).
32. Whitehead, A.N. Science and the Modern World *Cambridge University Press* (1925).
33. Whitehead, A.N. Process and Reality. *Cambridge University Press* (1929).
34. Penrose, R. Shadows of the Mind: A Search for the Missing Science of Consciousness. *Oxford University Press* (1989).
35. Hameroff, S.R. & Penrose, R. Orchestrated reduction of quantum coherence in brain microtubules: a model for consciousness. *Toward a science of consciousness; the first Tucson discussions and debates. Edited by S.R. Hameroff, A.W. Kaszniak, & A.C. Scott, MIT Press* (1996).

36. Tegmark, M. The importance of quantum decoherence in brain processes. *Physical Review* E61, 4194-4206 (2000).
37. Hameroff, S.R. & Penrose, R. Reply to seven commentaries on "Consciousness in the universe: Review of the 'Orch OR' theory". *Physics of Life Reviews* 11, 94-100 (2014).
38. Pockett, S. The Nature of Consciousness: A Hypothesis. *iUniverse Press* (2000).
39. McFadden, J. The Conscious Electromagnetic Information (Cemi) Field Theory: The Hard Problem Made Easy? *J Consciousness Studies* 9, 45-60 (2002).
40. McFadden, J. The CEMI Field Theory Gestalt Information and the Meaning of Meaning. *J Consciousness Studies* 20, 3-4 (2013).
41. Heusser, K., Tellschaft, D., & Thoss, F. Influence of an alternating 3 Hz magnetic field with an induction of 0.1 millitesla on chosen parameters of the human occipital EEG. *Neurosci Letters* 239, 57-60 (1997).
42. Carrubba, S., Frilot, C.; Chesson, A.L.; Webber, C.L.; Zbilut, J.P., & Marino, A.A. Magnetosensory evoked potentials: consistent nonlinear phenomena. *Neurosci Res* 60, 95-105 (2008).
43. Fröhlich, F. & McCormick, D.A. Endogenous Electric Fields May Guide Neocortical Network Activity. *Neuron* 67, 129–143 (2010).
44. Anastassiou, C.A., Perin, R., Markram, H., Koch, C. Ephatic coupling of cortical neurons. *Nat Neurosci* 14, 217-223 (2011).
45. Baars, B.J. & Edelman, D.B. Consciousness, biology and quantum hypotheses. *Physics of Life Reviews* 9, 285-294 (2012).
46. Edelman, G. Wider Than the Sky: The Phenomenal Gift of Consciousness. *Yale University Press* (2004).

47. Damásio, A. The Feeling of What Happens: Body and Emotion in the Making of Consciousness. *Mariner Books* (2000).

48. Baars, B.J. In the Theater of Consciousness: Global Workspace Theory. *J Consciousness Studies* 4, 292-309 (1997).

49. Dehaene, S. Consciousness and the brain: Deciphering How the Brain Codes Our Thoughts *Viking Press* (2014).

50. Gazzaniga, M.S. Who's in Charge?: Free Will and the Science of the Brain. *Ecco Press, New York* (2011).

51. Changeux, J.P. & Connes, A. Conversations on Mind, Matter and Mathematics. *Edited by M. B. DeBevoise, Princeton University Press* (1995).

52. Balduzzi, D. & Tononi, G. Qualia: The Geometry of Integrated Information. *PLOS Computational Biology*, 5(8): E1000462 (2009).

53. Tononi, G., Boly, M., Massimini, M., & Koch, C. Integrated information theory: from consciousness to its physical substrate. *Nat Rev Neurosci* 17, 450–461 (2016).

54. Seth, A.K. The grand challenge of consciousness. *Frontiers in Psychology* 1, 5 (2010).

55. Overgaard, M. The Status and Future of Consciousness Research. *Frontiers in Psychology* 8, 1719 (2017).

56. Bigelow, H.J. Dr. Harlow's Case of Recovery from the Passage of an Iron Bar through the Head. *American Journal of the Medical Sciences* 20, 13-22 (1850).

57. Harlow, J.M. Recovery from the passage of an iron bar through the head. *Publications of the Massachusetts Medical Society* 2, 1-22 (1868).

58. Kean, S. Phineas Gage: Neuroscience's Most Famous Patient. *Slate* (2014).

59. Larson, K.C. Rosemary: The Hidden Kennedy Daughter. *Mariner Books* (2015).

60. Hnasko, T.S., Perez, F.A., Scouras, A., Stoll, E.A., Luquet, S., Gale, S.D., Phillips, P.E., Kremer, E.J., & Palmiter, R.D. Cre-recombinase mediated restoration of dopamine production to the dorsal striatum results in hyperphagia and hyperactivity. . *Proceedings of the National Academy of Sciences USA* 103, 8858-8863 (2006).

61. Barbeau, A. L-Dopa Therapy in Parkinson's Disease: A Critical Review of Nine Years' Experience. *Can Med Assoc J.* 101, 59–68 (1969).

62. Monti, M.M., Vanhaudenhuyse, A., Coleman, M.R., Boly, M., Pickard, J.D., Tshibanda, L., Owen, A.M., & Laureys, S. Willful modulation of brain activity in disorders of consciousness. *New England Journal of Medicine* 362, 579-589 (2010).

63. Cruse, D., Chennu, S., Chatelle, C., Bekinschtein, T.A., Fernández-Espejo, D., Pickard, J.D., Laureys, S., & Owen, A.M. Bedside detection of awareness in the vegetative state: a cohort study. *Lancet* 278, 2088-2094 (2011).

64. Stoll, J., Chatelle, C., Carter, O., Koch, C., Laureys, S., & Einhäuser, W. Pupil responses allow communication in locked-in syndrome patients. *Current Biol* 23, R647-648 (2013).

65. Tuttle, A.H., Tohyama, S., Ramsay, T., Kimmelman, J., Schweinhardt, P., Bennett, G.J., & Mogil, J.S. Increasing placebo responses over time in U.S. clinical trials of neuropathic pain. *Pain* 156, 2616-2626 (2015).

66. Wechsler, M.E., Keller, J.M., Boyd, I.O., Dutile, S., Marigowda, G., Kirsch, I., Israel, E., & Kaptchuk, T.J. Active albuterol or placebo, sham acupuncture, or no intervention in asthma. *New England Journal of Medicine* 365, 119-126 (2011).

67. Merz, M., Seiberling, M., Hoxter, G., Holting, M., & Wortha, H. Elevation of liver enzymes in multiple dose trials during placebo treatment: Are they predictable? *J Clin Pharmacol* 37, 791-798 (1997).

68. Hashish, I., Hai, H.K., Harvey, W., Feinmann, C., & Harris, M. Reduction of post-operative pain and swelling by ultrasound: A placebo effect. *Pain* 33, 303-311 (1988).
69. Finniss, D.G., Kaptchuk, T.J., Miller, F. & Benedetti, F. Biological, clinical, and ethical advances of placebo effects. *Lancet* 375, 686-695 (2010).
70. Danese, S., Schabel, E., Masure, J., Plevy, S., & Schreiber, S. Are We Ready to Abandon Placebo in Randomised Clinical Trials for Inflammatory Bowel Disease? Pros and Cons. *J Crohns Colitis* Supp 2, S548-552 (2016).
71. Murray, D. & Stoessl, A.J. Mechanisms and therapeutic implications of the placebo effect in neurological and psychiatric conditions. *Pharmacology & Therapeutics* 140, 306-318 (2013).
72. Benedetti, F. Mechanisms of Placebo and Placebo-Related Effects Across Diseases and Treatments. *Annual Rev Pharmacology Toxicology* 48, 33-60 (2008).
73. Nietzsche, F. The Gay Science. *Translated with commentary by Walter Kaufmann in 1974, Vintage Books* (1882).
74. Kim, J. Physicalism, or Something Near Enough. *Princeton University Press* (2007).
75. Rochat, P. Five levels of self-awareness as they unfold early in life. *Consciousness and Cognition* 12, 717-731 (2003).
76. Cvencek, D., Greenwald, A.G., & Meltzoff, A.N. Implicit measures for preschool children confirm self-esteem's role in maintaining a balanced identity. *Journal of Experimental Social Psychology* 62, 50-57 (2016).
77. Perner, J., Leekam, S., & Wimmer, H. Three-year-olds' difficulty with false belief: The case for a conceptual deficit. *British Journal of Developmental Psychology* 5, 125-137 (1987).
78. Gopnik, A. & Astington, J.W. Children's understanding of representational change and its relation to the understanding of false belief and the appearance-reality distinction. *Child Development* 59, 26-37 (1988).

79. Wells, C., Morrison, C.M., & Conway, M.A. Adult recollections of childhood memories: What details can be recalled? *The Quarterly Journal of Experimental Psychology* 67, 1249-1261 (2014).
80. Demiray, B. & Bluck, S. The relation of the conceptual self to recent and distant autobiographical memories. *Memory* 19, 975-992 (2011).
81. Gibbs, J.W. Elementary Principles in Statistical Mechanics, developed with especial reference to the rational foundation of thermodynamics. *Charles Scribner's Sons* (1902).
82. Shannon, C.E. A Mathematical Theory of Communication. *Bell System Technical Journal* 27, 379–423 (1948).
83. von Neumann, J. Mathematical Foundations of Quantum Mechanics. *Springer* (1932).
84. Schiff, N.D., Nauvel, T., & Victor, J.D. Large-scale brain dynamics in disorders of consciousness. *Curr Opin Neurobiol* 25, 7-14 (2014).
85. Engel, A.K. & Singer, W. Temporal binding and the neural correlates of sensory awareness. *Trends Cogn Sci* 5, 16-25 (2001).
86. Harris, A.Z. & Gordon, J.A. Long-range neural synchrony in behavior. *Annu Rev Neurosci* 38, 171-194 (2015).
87. Grion, N., Akrami, A., Zuo, Y., Stella, F., & Diamond, M.E. Coherence between Rat Sensorimotor System and Hippocampus Is Enhanced during Tactile Discrimination. *PLoS Biol* 14, e1002384 (2016).
88. Goddard, C., Sridharan, D., Huguenard, J.R., & Knudsen, E. Gamma oscillations are generated locally in an attention-related midbrain network. *Neuron* 73, 567-580 (2012).
89. Tseng, P., Chang, Y.T., Chang, C.F., Liang, W.K. & Juan, C.H. The critical role of phase difference in gamma oscillation within the temporoparietal network for binding visual working memory. *Sci Rep* 6, 32138 (2016).

90. Sheridan, G.K., Moeendarbary, E., Pickering, M., O'Connor, J.J. & Murphy, K.J. Theta-burst stimulation of hippocampal slices induces network-level calcium oscillations and activates analogous gene transcription to spatial learning. *PLoS One* 9, e100546 (2014).
91. Zarnadze, S., Bauerle, P., Santos-Torres, J., Bohm, C., Schmitz, D., Geiger, J.R., Dugladze, T., & Gloveli, T. Cell-specific synaptic plasticity is induced by network oscillations. *Elife* 5 (2016).
92. Hebb, D.O. The Organization of Behavior. *Wiley & Sons* (1949).
93. Levy, W.B. & Steward, O. Temporal contiguity requirements for long-term associative potentiation and depression in the hippocampus. *Neuroscience* 8, 791–797 (1983).
94. Heisenberg, W. Über den anschaulichen Inhalt der quantentheoretischen Kinematik und Mechanik. *Zeitschrift für Physik* 43, 172–198 (1927).
95. Schrödinger, E. An Undulatory Theory of the Mechanics of Atoms and Molecules. *Physical Review* 28, 1049–1070 (1926).
96. Jammer, M. The philosophy of quantum mechanics. *Wiley & Sons, Inc.* (1974).
97. Kudaka, S. & Matsumoto, S. . Uncertainty principle for proper time and mass. *J Math Phys* 40, 1237-1245 (1999).
98. Feynman, R.P. Forces in Molecules. *Physical Review* 56, 340 (1939).
99. Kaluza, T. Zum Unitätsproblem in der Physik. *Sitzungsberichte der Preussischen Akademie der Wissenschaften (Math/Phys)*, 966–972 (1921).
100. Klein, O. Quantentheorie und fünfdimensionale Relativitätstheorie. *Zeitschrift für Physik* 37, 895–906 (1926).
101. Dorval, A.D. & White, J.A. Channel noise is essential for perithreshold oscillations in entorhinal stellate neurons. *J Neurosci* 25, 10025-10028 (2005).

102. Stern, E.A., Kincaid, A.E., & Wilson, C.J. Spontaneous subthreshold membrane potential fluctuations and action potential variability of rat corticostriatal and striatal neurons in vivo. *J Neurophysiol* 77, 1697-1715 (1997).

103. Softky, W.R. & Koch, C. The highly irregular firing of cortical cells is inconsistent with temporal integration of random EPSPs. *J Neurosci* 13, 334-350 (1993).

104. Powers, R.K. & Binder, M.D. Effective synaptic current and motoneuron firing rate modulation. *J Neurophysiol* 74, 793-801 (1995).

105. Bialek, W. & Rieke, F. Reliability and information transmission in spiking neurons. *Trends Neurosci* 15, 428-434 (1992).

106. Fairhall, A.L., Lewen, G.D., Bialek, W., & de Ruyter van Steveninck, R.R. Efficiency and ambiguity in an adaptive neural code. *Nature* 412, 787-792 (2001).

107. Edwards, D.H., Yeh, S.R., & Krasne, F.B. Neuronal coincidence detection by voltage-sensitive electrical synapses. *Proc Natl Acad Sci U S A* 95, 7145-7150 (1998).

108. Stacey, W.C., Krieger, A., & Litt, B. Network recruitment to coherent oscillations in a hippocampal computer model. *J Neurophysiol* 105, 1464-1481 (2011).

109. Thiele, A. & Stoner, G. Neuronal synchrony does not correlate with motion coherence in cortical area MT. *Nature* 421, 366-370 (2003).

110. Herrmann, C.S., Munk, M.H., & Engel, A.K. Cognitive functions of gamma-band activity: memory match and utilization. *Trends Cogn Sci* 8, 347-355 (2004).

111. Pulvermuller, F., *et al.* High-frequency cortical responses reflect lexical processing: an MEG study. *Electroencephalogr Clin Neurophysiol* 98, 76-85 (1996).

112. Rodriguez, E., *et al.* Perception's shadow: long-distance synchronization of human brain activity. *Nature* 397, 430-433 (1999).

113. Howarth, C., Gleeson, P., & Attwell, D. Updated energy budgets for neural computation in the neocortex and cerebellum. *J Cereb Blood Flow Metab* 32, 1222-1232 (2012).
114. Engl, E. & Attwell, D. Non-signalling energy use in the brain. *J Physiol* 593, 3417-3429 (2015).
115. Grandy, W.T. Entropy and the time evolution of macroscopic systems. *Oxford University Press* (2008).
116. Hillert, M. & Agren, J. Extremum principles for irreversible processes. *Acta Materialia* 54, 2063-2066 (2006).
117. Martyushev, L.M., Nazarova, A.S., & Seleznev, V.D. On the problem of minimum entropy production in the nonequilibrium stationary state. *Journal of Physics A: Mathematical and Theoretical* 40, 371-380 (2007).
118. Still, S., Sivak, D.A., Bell, A.J. & Crooks, G.E. Thermodynamics of prediction. *Physical Review Letters* 109, 120604 (2012).
119. Kok, P., Brouwer, G.J., van Gerven, M.A. & de Lange, F.P. Prior expectations bias sensory representations in visual cortex. *J Neurosci* 33, 16275-16284 (2013).
120. Kok, P., Jehee, J.F. & de Lange, F.P. Less is more: expectation sharpens representations in the primary visual cortex. *Neuron* 75, 265-270 (2012).
121. Denisyuk, Y.N. Fundamentals of Holography. *Translated into English by A. Chubarov, Mir Publishers* (1978).
122. Caulfield, H.J. & Shamir, J. Holograms of real and virtual point trajectories. In: Three-dimensional holographic imaging, Chapter 2. *Wiley & Sons* (2002).
123. Ross, M.P. & Shumlak, U. Digital holographic interferometry employing Fresnel transform reconstruction for the study of flow shear stabilized Z-pinch plasmas. *Rev Sci Instrum* 87, 103502 (2016).
124. Lawrence, J.R., O'Neill, F.T., & Sheridan, J.T. Photopolymer holographic recording material. *Optik* 112, 449-463 (2001).

125. Insanally, M.N., *et al.* Spike-timing-dependent ensemble encoding by non-classically responsive cortical neurons. *Elife* 8 (2019).

126. Roxin, A., Brunel, N., Hansel, D., Mongillo, G., & van Vreeswijk, C. On the distribution of firing rates in networks of cortical neurons. *J Neurosci* 31, 16217-16226 (2011).

127. Steinmetz, P.N., Manwani, A., Koch, C., London, M., & Segev, I. Subthreshold voltage noise due to channel fluctuations in active neuronal membranes. *J Comput Neurosci* 9, 133-148 (2000).

128. Gabor, D. A new microscopic principle. *Nature* 161, 777 (1948).

129. Mouritsen, O.G. Self-assembly and organization of lipid-protein membranes. *Current Opinion in Colloid & Interface Science* 3, 78-87 (1998).

130. Colburn, W.S. & Haines, K.A. Volume hologram formation in photopolymer materials. *Appl Opt* 10, 1636-1641 (1971).

131. Candelas, P. Complete intersection Calabi-Yau manifolds. *Nuclear Physics B* 298, 493-525 (1988).

132. Green, P.S. & Hübsch, T. Connecting moduli spaces of Calabi-Yau threefolds. *Communications in Mathematical Physics* 119, 431–441 (1988).

133. Berut, A., *et al.* Experimental verification of Landauer's principle linking information and thermodynamics. *Nature* 483, 187-189 (2012).

134. Jun, Y., *et al.* High-precision test of Landauer's principle in a feedback trap. *Phys Rev Lett* 113, 190601 (2014).

135. Yan, L.L., *et al.* Single-atom demonstration of the quantum Landauer principle. *Phys Rev Lett* 120, 210601 (2018).

136. Landauer, R. Irreversibility and heat generation in the computing process. *IBM Journal of Research and Development* 5, 183-191 (1961)

137. Steriade, M., Timofeev, I., & Grenier, F. Natural waking and sleep states: a view from inside neocortical neurons. *J Neurophysiol* 85, 1969-1985 (2001).
138. Collell, G. & Fauquet, J. Brain activity and cognition: a connection from thermodynamics and information theory. *Front Psychol* 6, 818 (2015).
139. Street, S. Neurobiology as Information Physics. *Front Syst Neurosci* 10, 90 (2016).
140. Bellec, G., *et al.* A solution to the learning dilemma for recurrent networks of spiking neurons. *Nat Commun* 11, 3625 (2020).
141. Keyes, D.E., Ltaief, H., & Turkiyyah, G. Hierarchical algorithms on hierarchical architectures. *Philos Trans A Math Phys Eng Sci* 378, 20190055 (2020).
142. Stoll, E.A. Thermodynamic computing in cortical neural networks. *Under review* (a).
143. Stoll, E.A. The mechanics underpinning non-deterministic computation in cortical neural networks. *Under review* (b).
144. Stoll, E.A. Modeling the probabilistic behavior of electrons at the neuronal membrane yields a holographic projection of information content. *Under review* (c).
145. Aaronson, S. Quantum computing, postselection, and probabilistic polynomial-time. *Proceedings: Mathematical, Physical and Engineering Sciences* 461, 3473-3482 (2005).
146. Preskill, J. Lecture Notes for Physics 229: Quantum Information and Computation. *California Institute of Technology* (1998).
147. Holevo, A.S. Bounds for the quantity of information transmitted by a quantum communication channel. *Probl Peredachi Inform* 9, 3-11 (1973).
148. Schumacher, B. Quantum coding. *Phys Rev A* 51, 2738 (1995).

149. Giustina, M., et al. Bell violation using entangled photons without the fair-sampling assumption. *Nature* 497, 227-230 (2013).
150. Henson, B., et al. Loophole-free Bell inequality violation using electron spins separated by 1.3 kilometres. *Nature* 526, pages 682–686 (2015).
151. Renou, M.O., et al. Quantum theory based on real numbers can be experimentally falsified. *Nature* 600, 625-629 (2021).
152. Solomon, S.G. & Lennie, P. The machinery of colour vision. *Nat Rev Neurosci* 8, 276-286 (2007).
153. Zhou, H., et al. Spatiotemporal dynamics of brightness coding in human visual cortex revealed by the temporal context effect. *Neuroimage* 205, 116277 (2020).
154. Swerdlow, N.R., Blumenthal, T.D., Sutherland, A.N., Weber, E., & Talledo, J.A. Effects of prepulse intensity, duration, and bandwidth on perceived intensity of startling acoustic stimuli. *Biol Psychol* 74, 389-395 (2007).
155. Chambers, J.D., et al. Computational Neural Modeling of Auditory Cortical Receptive Fields. *Front Comput Neurosci* 13, 28 (2019).
156. Pins, D. & Ffytche, D. The neural correlates of conscious vision. *Cereb Cortex* 13, 461-474 (2003).
157. Snyder, J.S., Schwiedrzik, C.M., Vitela, A.D., & Melloni, L. How previous experience shapes perception in different sensory modalities. *Front Hum Neurosci* 9, 594 (2015).
158. Mostert, P., Kok, P., & de Lange, F.P. Dissociating sensory from decision processes in human perceptual decision making. *Sci Rep* 5, 18253 (2015).
159. Hanks, T.D., Mazurek, M.E., Kiani, R., Hopp, E., & Shadlen, M.N. Elapsed decision time affects the weighting of prior probability in a perceptual decision task. *J Neurosci* 31, 6339-6352 (2011).

160. Bekenstein, J.D. Universal upper bound on the entropy-to-energy ratio for bounded systems. *Physical Review D* 23, 287 (1981).
161. Buzsaki, G. & Draguhn, A. Neuronal oscillations in cortical networks. *Science* 304, 1926-1929 (2004).
162. Pauli, W. Über den Zusammenhang des Abschlusses der Elektronengruppen im Atom mit der Komplexstruktur der Spektren. *Zeitschrift für Physik* 31, 765-783 (1925).
163. Stoll, E.A. The explanatory power of Conifold Theory. *Under review* (d).
164. Stoll, E.A. The neuroscientific predictions of Conifold Theory. *Under review* (e).
165. Hoffman, D. The case against reality: Why evolution hid the truth from our eyes. . *W.W. Norton & Company* (2019).
166. Brockman, H.L., Momsen, M.M., King, W.C., & Glomset, J.A. Structural determinants of the packing and electrostatic behavior of unsaturated phosphoglycerides. *Biophys J* 93, 3491-3503 (2007).
167. Islas, L.D. & Sigworth, F.J. Voltage sensitivity and gating charge in Shaker and Shab family potassium channels. *Journal of General Physiology* 114, 723-742 (1999).
168. Nicholson, A.A., et al. The neurobiology of emotion regulation in posttraumatic stress disorder: Amygdala downregulation via real-time fMRI neurofeedback. *Human Brain Mapp* 38, 541-560 (2017).
169. Carroll, K.M. et al. Computer-assisted delivery of cognitive-behavioral therapy for addiction: a randomized trial of CBT4CBT. *Am J Psych* 165, 881-888 (2008).
170. van der Koch, B. The Body Keeps the Score: Brain, Mind, and Body in the Healing of Trauma. *Penguin Books* (2014).
171. Gruene, T.M., Flick, K., Stefano, A., Shea, S.D., & Shansky, R.M. Sexually divergent expression of active and passive conditioned fear responses in rats. *Elife* 4, pii: e11352 (2015).

172. Wang, M.E., Wann, E.G., Yuan, R.K., Ramos-Álvarez, M.M., Stead, S.M., & Muzzio, I.A. Long-term stabilization of place cell remapping produced by a fearful experience. *J Neurosci* 32, 15802-15814 (2012).
173. Lee, K.A., Vaillant, G.E., Torrey, W.C., & Elder, G.H. A 50-year prospective study of the psychological sequelae of World War II combat. *Am J Psych* 152, 516-522 (1995).
174. McLaughlin, K.A., Sheridan, M.A., Gold, A.L., Duys, A., Lambert, H.K., Peverill, M., Heleniak, C., Shechner, T., Wojcieszak, Z., & Pine, D.S. Maltreatment Exposure, Brain Structure, and Fear Conditioning in Children and Adolescents. *Neuropsychopharmacology* 41, 1956-1964 (2016).
175. Wallace, D.F. Fate, Time, and Language: An Essay on Free Will. *Edited by J. Ryerson, Columbia University Press* (2010).

Acknowledgements

I would like to thank the colleagues who provided expert advice during the preparation of this manuscript and its new edition, especially Roman Bauer, Roozbeh Kiani, Alberto Elduque, Angus Nisbet, Bharat Ratra, Ian Durham, Alex Nugent, and Max Aldana. Your time and patience in challenging my conceptual framework and pointing out errors vastly improved the manuscript at every stage. I would also like to thank Mike Shadlen, who set me on this path many years ago by assuring me that investigating neuronal processes as a function of time was the way forward for both neuroscience and my own career. Many thanks are also in order to Andrew Jackson for all the interesting conversations about neurophysiology and physics over the years.

I am deeply grateful to the friends who have always shown me immense kindness, love, and support, especially Amy Farrar, Dustin Key, Thomaie Hilaris, Ryan Neris, Deb and Tim Guirl, Kristina and Philip Tilker, Nora Kochie, Samantha Slater, Lilia Lim, Hyon Rah, Joy Wattawa, Deb Hamilton, and Jay Weiner.

I also wish to express my gratitude to my many mentors, colleagues, and friends in neuroscience, who over the years (and beers) gave me so much to ponder, especially Michael Gabriel, the first person to give me a chance at working in a lab; Jeff Mogil, who sent me in the right direction; Jenn Norton, my first friend in science; Andrew Talk, my first mentor; Lauren Burhans and Amir Kashef, who always had words of wisdom to share; Chris Spatz, for teaching me thermodynamics; Talley Lambert, for pointing out the implications of miniature post-synaptic potentials on neuronal coincidence detection; Andrew Gartland, for being sensible about everything but strawberries; Abby and Barry Wark, for proving that you can do several things at once and do them well; Minhui Lee, Christy Walcher, Kevin Curran, Richard Row, and Barbara Wakimoto, for teaching me about developmental biology; Lidong Liu, for teaching me mad skills; Rian de Laat and Megan Hamner, for telling the truth; Rheba de Tornyay, for whispering those words in my ear; Phil Horner, for all the advice that I only understood later; Robert Rostomily, for the constant inspiration; Mark Bothwell, Richard Palmiter, and Tom Reh, for the conversations which helped me choose my path; Robert Steiner, for providing a solid ground; Bertil Hille, for being patient with all of us; Ian Sweet, for teaching me metabolic physiology; Kristin Swanson, for making math seem okay again; Olivier Saut, for giving me a boost of confidence; Doug Turnbull, for the funding and the freedom; Chris Faulkes and Ross Maxwell, for the productive collaborations; Carolina Gandara and Glenda Watson, for the learning opportunities; Jeri Wright, for holding my hand; Denise Inman, for the smiles and support over the years; Paige Cundiff, for teaching me to take risks and trust myself; Llwyd Orton and Faiza Ben Mabrouk, for the giggles; Berit Jacobsen, for the ice cream timing trick; Jurate Lasiene, for being there to celebrate no matter which continent the party's on; Anne Gruenewald and Karolina Rygiel, for the perspective when I needed it; Fabio Gualtieri, for the coffee when I needed it; Bas Olthof, for the sympathy when I needed it; and Valerie Affleck, for keeping the faith as long as possible.

And finally, I would like to thank my students Benny Habibi, Willy Cheung, Nevena Karapavlovic, Michael Woodmass, Rebecca Makin, Katie Sung, James Bryson, Claire Boothman, Kevin Ho, Qasim Majid, Tim Porter, Shaan Patel, Paul Tang, Jonathan Stockton, Hannah Woodward, Gian Montevecchio, Emma Hall, Marie Strickland, Lucy Gee, Maggie Adiamah and Hua Lin; you each showed wisdom beyond your years.

www.ingramcontent.com/pod-product-compliance
Lightning Source LLC
Chambersburg PA
CBHW030305080526
44584CB00012B/447